Vascular Smooth Muscle

Der Gefäßmuskel

Proceedings of the Satellite-Symposium of the XXV. International Congress of Physiological Sciences and Annual Meeting of the German Angiological Society July 20–24, 1971 in Tübingen

Verhandlungen des Satellit-Symposiums des XXV. Internationalen Kongresses für Physiologische Wissenschaften und der Jahrestagung der Deutschen Gesellschaft für Angiologie e. V.
Vom 20.–24. Juli 1971 in Tübingen

Editor: E. Betz

Mit 40 Abbildungen

Springer-Verlag Berlin Heidelberg New York 1972

ISBN 3-540-05711-0 Springer-Verlag Berlin Heidelberg New York
ISBN 0-387-05711-0 Springer-Verlag New York Heidelberg Berlin

Foreword

This book summarizes the papers presented at the International Symposium on vascular smooth muscle, organized in combination with the XXV International Congress of Physiological Sciences by the German Society of Angiology e. V. Honorary Presidents were Professor Dr. Dr. h. c. K. Wezler and Professor Dr. Dr. h. c. H. E. Bock. The annual meeting of the Society of Angiology was combined with the Symposium. It was held at the Physiological Institute of the University of TÜBINGEN, an appropriate setting, since in the last century numerous scientists from this institute were engaged in special problems connected with the vascular system and muscles.

The topics of this meeting covered a wide field of vascular research; it was the aim of the organizers to the discuss the physiological, biochemical and morphological basis of the vascular muscles with clinicians and to stress some important aspects in the form of round table discussions.

On the following pages the reader will therefore find numerous papers on special basic problems of the physiology, biochemistry and morphology of vascular smooth muscle as well as on pathophysiological and pathological changes and papers about the clinical treatment of vascular disorders. The more experimental and theoretical papers and discussions are concerned with the single smooth muscle cell and the different types of vessels in various organs whereas the discussion of the clinical sections concentrate on the coronary system.

Since so many papers were included, and in the interest of rapid publication, the essentials of the papers were summarized by the participants and were published together with discussion remarks summarized by the chairmen of the sections. The organizers wish to express their thanks to all who enlivened the meeting by their comments and enquiries.

The costs of the publication have been met by a generous grant from the Erwin-Riesch-Stiftung, Lorch/Württ., which is gratefully acknowledged. Dr. H. Schroeder and N. Fromann, Fa. Dr. Karl Thomae, Biberach/Riß were all of great help in organizing the conference. Assistance towards the costs of the Symposium was also received from the following bodies, to whom thanks are recorded:

Byk-Gulden Lomberg GmbH., Konstanz
Chemische Werke Albert, Wiesbaden-Biebrich
Chemiewerke Homburg, Frankfurt/M.
Ciba AG., Wehr/Baden
Deutsche Farmitalia GmbH., Freiburg i. B.
Getränke-Gesellschaft Max Schmeling u. Co. KG., Reutlingen
Idee-Kaffee, J. J. Darboven, Hamburg

Janssen GmbH., Düsseldorf
Knoll AG., Chemische Fabriken, Ludwigshafen/Rh.
Luitpold-Werk, Chem.-Pharmazeutische Fabrik, München
E. Merck, Darmstadt, Moorheilbad Bad Buchau gGmbH
Pfizer GmbH., Karlsruhe
J. Pfrimmer u. Co., Pharmazeutische Werke, Erlangen
Sandoz AG., Nürnberg
Dr. Wilmar Schwabe, Karlsruhe
Johann A. Wülfing-Bauer u. Cie., Düsseldorf

Vorwort

Dieses Buch enthält eine Zusammenfassung eines internationalen Symposiums
über Gefäßmuskulatur anläßlich des XXV. Internationalen Kongresses für
Physiologische Wissenschaften und wurde durch die Deutsche Gesellschaft
für Angiologie e.V. unter dem Ehrenvorsitz von Professor Dr. Dr. h.c.
K. Wezler und Professor Dr. Dr. h.c. H.E. Bock veranstaltet. Die Gesellschaft verband mit dem Symposium ihre Jahrestagung. Die Tagung fand im
Physiologischen Institut der Universität Tübingen statt, einem Institut, das
sich seit einem Jahrhundert mit vielen speziellen Problemen des Gefäßsystems und der Muskulatur beschäftigt. Die Thematik umspannte einen großen
Bereich der Gefäßforschung; es war ein besonderes Anliegen der Tagungs-
leitung, physiologische, biochemische und morphologische Grundlagen mit
Klinikern zu diskutieren sowie für die klinische Praxis wichtige Aspekte in
Form von Rundtischgesprächen hervorzuheben. Auf den folgenden Seiten findet der Leser daher eine Reihe von Vortragszusammenfassungen, die die
spezielle Grundlagenforschung auf dem Gebiet der Physiologie, Biochemie
und Morphologie des Gefäßmuskels zum Thema haben, neben solchen, die
sich mit pathophysiologischen und pathologischen Veränderungen sowie klinischen und therapeutischen Anwendungen befassen. Während die mehr experimentell und theoretisch ausgerichteten Vorträge und Diskussionen sich mit
der einzelnen glatten Muskelzelle der verschiedenartigsten Gefäßgebiete
beschäftigten, waren die Vorträge aus der angewandten Physiologie auf
spezielle Bedingungen an verschiedenen Organen ausgerichtet. Der Schwerpunkt der Vorträge aus dem Bereich der klinischen Disziplinen betraf das
Coronargefäßsystem.

Im Interesse einer schnellen Veröffentlichung und deshalb, weil so sehr viele
Vorträge eingereicht wurden, war es nicht möglich, die Vorträge vollständig
zu veröffentlichen. Die Autoren haben vielmehr die wichtigsten Gesichts-
punkte zusammengefaßt und diese zur Publikation eingereicht. Die Diskussionen konnten aus den gleichen Gründen ebenfalls nicht in der gesamten Breite veröffentlicht werden. Sie wurden von den jeweiligen Vorsitzenden und
ihren Mit-Vorsitzenden zusammengefaßt und in Form von Diskussionszusam-
menfassungen veröffentlicht. Das gleiche gilt für die Rundtischgespräche.
Die Organisatoren bedanken sich bei allen, die die Tagung durch ihre Kom-
mentare und Fragen belebt haben. Die Kosten der Veröffentlichung wurden
durch eine großzügige Spende der Erwin-Riesch-Stiftung, Lorch/Württ. ermöglicht, für die hier Dank gesagt werden soll. Bei der Organisation der
Veröffentlichung und des Symposiums waren uns die Herren Dr. H. Schroeder
und N. Frommann, Firma Dr. Karl Thomae, Biberach/Riß eine wertvolle
Hilfe. Die Durchführung des Symposiums wurde durch Spenden folgender
Firmen - denen hier sehr herzlich gedankt werden soll - ermöglicht:

Byk-Gulden Lomberg GmbH, Konstanz
Chemische Werke Albert, Wiesbaden-Biebrich
Chemiewerk Homburg, Frankfurt/M.
Ciba AG., Wehr/Baden
Deutsche Farmitalia GmbH., Freiburg i.B.
Getränke-Gesellschaft Max Schmeling u. Co. KG.   Reutlingen
Idee-Kaffee, J.J. Darboven, Hamburg
Janssen GmbH., Düsseldorf
Knoll AG., Chemische Fabriken, Ludwigshafen/Rh.
Luitpold-Werk, Chem.-Pharmazeut. Fabrik, München
E. Merck, Darmstadt, Moorheilbad Bad Buchau gGmbH
Pfizer GmbH, Karlsruhe
J. Pfrimmer u. Co., Pharmazeutische Werke, Erlangen
Sandoz AG., Nürnberg
Dr. Wilmar Schwabe, Karlsruhe
Johann A. Wülfing-Bauer u. Cie., Düsseldorf

Contents/Inhalt

Session 5: Session 4 continued: Specific ionic action and
          osmotic effects.
Sitzung 5: Fortsetzung der Sitzung 4 sowie spezifische Ionen-
           wirkungen und osmotische Effekte.

Contributors/Mitarbeiter

Electrophysiological Studies on Spontaneous Activity of Vascular Smooth Muscle

By K. Golenhofen
Physiologisches Institut der Universität Marburg/Lahn, Germany

A classification of smooth muscle rhythms is derived from comparative studies in different smooth muscle preparations. On this basis, the spontaneous activity of the portal vein - as an example of a blood vessel with highly pronounced spontaneous activity - can be described as a superimposition of several rhythms: 1) minute-rhythm (MR), which has a frequency of 0.5-2/min and is combined with only small fluctuations in the basic membrane potential; 2) basic organ specific rhythm (BOR) which divides the active phases of the MR and is connected with potential fluctuations of 10-30 mV; its frequency is 5-10 min;
3) faster oscillations of the membrane potential with period durations of ca. 1 sec (second-rhythm, SR);
4) spikes with amplitudes of between 40 and 60 mV, triggered by the SR and acting as a signal to trigger the mechanical tension development (GOLENHOFEN and v. LOH, 1970). Spikes and SR of portal vein are similar in mechanism to those of intestinal smooth muscle: They are, in the same way, Ca-dependent and are suppressed by iproveratril (GOLENHOFEN and LAMMEL, 1971). The MR is also similar in the different tissues. It therefore appears permissible to take smooth muscle of portal vein as the functional basis of vascular smooth muscle in general. Other types of vascular smooth muscle can be described as having developed from this basis by specialization and differentiation. The order of specialization can be seen - as a working hypothesis - roughly as follows: Portal vein - terminal venous vessels - terminal arterial resistance vessels - larger arteries - great elastic arteries. Myogenic automaticity is suppressed in this order, nervous control becomes stronger and more specialized and other processes appear in adaptation to special functions. The special properties of the action potential in elastic arteries can also be interpreted as part of a continuous system of vascular smooth muscle. As shown in fig. 1, the action potentials of elastic arteries are more similar in shape to the SR oscillation of portal vein and taenia coli than to the spike component of these tissues (the electrical activity of taenia coli in Ca-free solution, in fig. 1c, can be described as SR oscillations without the Ca-dependent spike component), and they, as the SR are more Na-dependent (KEATINGE, 1968).

References

1) GOLENHOFEN, K., LAMMEL, E.: Pflügers Arch. (in press).
2) GOLENHOFEN, K., v LOH, D.: Pflügers Arch. 319, 82-100 (1970).
3) KEATINGE, W.R.: J. Physiol. 194, 169-182 (1968).

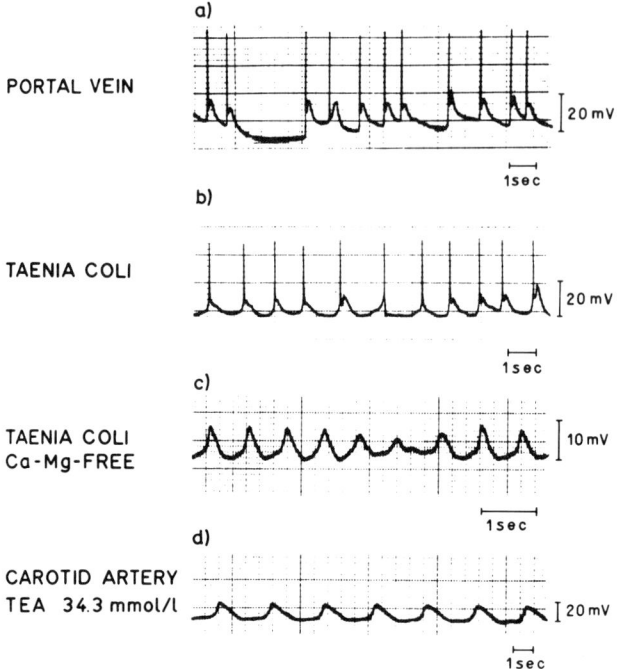

a)

PORTAL VEIN

b)

TAENIA COLI

c)

TAENIA COLI
Ca-Mg-FREE

d)

CAROTID ARTERY
TEA 34.3 mmol/l

Normal electrical activity of portal vein (a) and taenia coli (b) of the guinea-pig, consisting of SR oscillations and spikes, compared with the electrical activity of taenia coli in Ca-Mg-free solution (SR only, part c) and action potentials of rabbit common carotid artery in hypertonic solution containing TEA 34.3 m mol/l (part d, redrawn from MEKATA, J. gen. Physiol., in press). Membrane potential always measured intracellularly. The arterial action potentials (d) are comparable to the SR oscillations of the other sections.

Slow Oscillations of Transmembrane Na und K Fluxes in Vascular Smooth Muscle

By G. Siegel, H. P. Koepchen, and H. Roedel
Physiologisches Institut der Freien Universität Berlin, Germany

Vascular smooth muscle possesses all of the properties of a rhythmic system (3). Spontaneous, rhythmic contractions of blood vessels occur, parallel to spikes and oscillations of the membrane potential. These phenomena must be based on corresponding ionic movements which have not yet been observed.

Na^+ and K^+ exchange measurements at intervals Δ t of 18 sec were performed on isolated carotid media of dogs by use of a double tracer technique with $^{24}_{11} Na^+$ and $^{42}_{19} K^+$ (1,4). Ideal washout curves were fitted by a multiple-exponential approximation with the aid of a CDC 3300 digital computer. The time course of the deviations of measured values from the fitted curves was autocorrelated, and power spectra were calculated (6).

After prewhitening the uniformly spaced (Δ t) values Δ_t :

$$\Delta_t := \Delta_t - 0.6 \cdot \Delta_{t-1} \tag{1}$$

which corresponds to a multiplication of the Fourier-transform P (v) by $1.36 - 1.20 \cos 2$, the autocorrelation function was calculated

$$\varphi (\tau = r \Delta t) = \frac{1}{n - r} \sum_{q=0}^{n-r} \Delta_q \cdot \Delta_{q+r} - (\frac{1}{n} \sum_{q=0}^{n} \Delta_q)^2 , \tag{2}$$

where is n = duration of record in units of Δ t and r = duration of lag (τ) in units of Δ t. Finally, after correction for prewhitening, the finite cosine series transform of this function φ ($\tau = r \Delta$ t) is the power spectrum

$$\phi (v) = 2 \int_0^\infty \varphi (\tau) \cdot \cos \omega \tau \, d \tau . \tag{3}$$

By means of such an analysis a very distinct periodicity can be demonstrated in the Na^+ efflux as well as in the K^+ efflux. A typical example of a simultaneous $^{24}_{11} Na$ - $^{42}_{19} K$ efflux experiment using indirect methods in normal Krebs solution is shown in fig. 1. Both efflux curves illustrate the time course of the instantaneous exchange quotients. The ideal time course is given by the thinly drawn, computer fitted curves. In the initial, steep part of the washout curves, the flux oscillations are much smaller than in the later, flat course. This indicates that they are related to the cellular exchange. This behaviour is clarified in the insets labelled a in fig. 1 in which the percent

We thank Mrs. A. IDEN for her excellent technical assistance. This work was supported by the Deutsche Forschungsgemeinschaft and the Stiftung Volkswagenwerk.

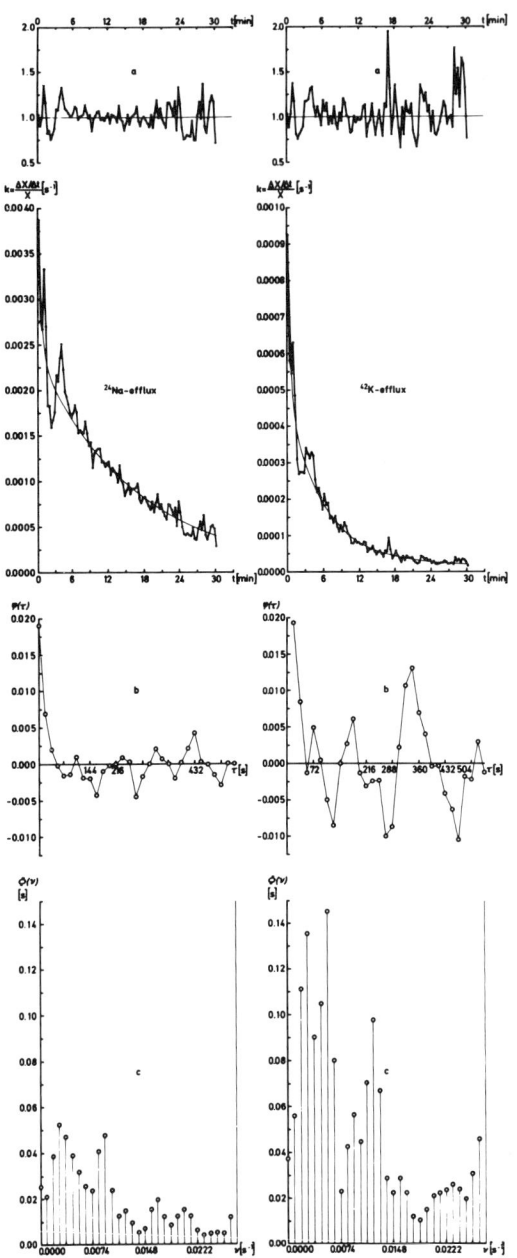

$^{24}_{11}$Na$^+$ and $^{42}_{19}$K$^+$ efflux determined simultaneously in the dogs carotid media according to the indirect method. The graphs show the time course of the instantaneous exchange quotient k(t) in a single experiment. The thin-traced line represents the optimal triple-exponential function found by computer fitting. Inset a: Relative deviation of the measured values from the functional values of the optimal curve. Inset b: Autocorrelation functions of the percent deviations given in part a. Oscillations are already clearly recognizable. Inset c: Power spectra of the autocorrelation functions. The frequencies of the oscillations shown in b may be directly observed on the abscissa of the power spectra.

deviation Δ_t of the measured values from the function values of the fitted
flux curves are given. Increase and decrease rates for the Na and K effluxes
of 20-100 % are obtained. Parts b and c of fig. 1 result from calculation of
the autocorrelation functions and the power spectra using the deviation, Δ_t.
Flux oscillations are already clearly seen in the autocorrelations. The power
spectrum enables an analysis of the frequencies of the oscillations. The frequency with the greatest amplitude is also the most probable one. From part
c of fig. 1 oscillations with period durations of 360, 108, 60 and 49 s for Na^+
and 180, 360, 83 and 108 s for K^+ have been ranked according to probability.

If one performs this analytical procedure on a series of experiments, flux
oscillations are obtained with periods between 39 s and 9 min. The analysis
shows that in all experiments oscillations of the Na^+ efflux appear more frequently than of the K^+ efflux (6). The principal period length for the Na^+ efflux
is 83 s and for the K^+ efflux 180 s.

Fluctuations in Na^+ efflux are probably due to periodic changes in active
transport, since in this tissue this component amounts to about 80 % of the
total Na^+ efflux. It seems appropriate to invoke an oscillatory, electrogenic
Na pump (5). On the other hand, the oscillating K^+ efflux could be interpreted as a consequence of an active transport mechanism. In this case, however, it is not certain if the oscillations in K^+ exchange result from an electrogenic, oscillating, inwardly directed K pump or arise as a result of slow
waves in the membrane potential produced by an oscillating, electrogenic Na
pump. The various oscillations in an individual preparation can be attributed
to differing cell fractions or to the presence of several pacemakers.

Since active transport is dependent on metabolism, the explanation for these
phenomena should at first be sought in rhythmic metabolic processes, e.g.
oscillations of glycolysis or mitochondrial events. Using suitable assump-
tions flux oscillations can be simulated by a mathematical model (2) of membrane is produced by an approximation of the Na and K permeabilities, such
that flux and potential oscillations occur.

References

1) KOEPCHEN, H.P., SIEGEL, G., WARTA, H.: Ionengehalt und Fluxe an
der glatten Gefäßmuskulatur. Pflügers Arch. ges. Physiol. 297, R 63
(1967).
2) ROEDEL, H., SIEGEL, G.: Simulation of membrane properties. Symposium "Passive permeability of cell membrane", Rotterdam, Netherlands.
July 20-22, 1971.
3) SELLER, H., LANGHORST, P., POLSTER, J., KOEPCHEN, H.P.: Zeitliche Eigenschaften der Vasomotorik. II. Erscheinungsformen und Entstehung spontaner und nervös induzierter Gefäßrhythmen. Pflügers Arch.
ges. Physiol. 296, 110 (1967).
4) SIEGEL, G., KOEPCHEN, H.P., ROEDEL, H.: Zur Bedeutung der K-
und Na-Ionen für das Ruhepotential der glatten Gefäßmuskulatur. Pflügers
Arch. ges. Physiol. 297, R 64 (1967).

5) SIEGEL, G., KOEPCHEN, H. P., ROEDEL, H., SCHOTT, A., MÜLLER, W. A.; with the technical assistance of LINZNER, U.: Oscillating fluxes and membrane potentials in vascular smooth muscle and cardiac muscle. Pflügers Arch. ges. Physiol. (in preparation).

6) SIEGEL, G., ROEDEL, H., KOEPCHEN, H. P.: Membrane permeability and active transport in vascular smooth muscle. I. European Biophysics Congress, Baden near Vienna, September 14-17, 1971.

Korrelation zwischen rhythmischen Spannungsänderungen der glatten Gefäßmuskulatur und der extra- und/oder intrazellulär abgeleiteten elektrischen Aktivität

Von G. Biamino und H.-J. Wessel
Medizinische Klinik und Poliklinik und Institut für Klinische Physiologie im Klinikum Steglitz der Freien Universität Berlin, Germany

Mit einer früher angegebenen Methode (Life Science 8, 157-162 (1969)) konnte gezeigt werden, daß an Aortenstreifenpräparaten zwischen der von zwei getrennten Stellen extrazellulär abgeleiteten elektrischen Aktivität und den in Abhängigkeit von Elektrolytänderungen oder Zugabe von vasoaktiven Substanzen simultan registrierten Spannungsänderungen stets eine strenge Korrelation besteht. Als bestimmende Parameter für die induzierten Spannungsänderungen kristallisierten sich einerseits die Effektivität der Erregungsüberleitung und andererseits Zahl, Regelmäßigkeit und Frequenz der aktiven Schrittmacherzentren heraus. Da jedoch die angewandte Paraffinmethode nur indirekt eine Aussage über Membranpotentialänderungen erlaubt, wurde sowohl von Aortenstreifenpräparaten, wie z. T. von Präparaten der Vena portae der Ratte in 25 Experimenten simultan neben der mechanischen die elektrische Aktivität mit der Mikroelektrodentechnik (eine intrazelluläre und drei extrazelluläre Elektroden) registriert und ihre Abhängigkeit von Bademilieuänderungen vor allem mit dem Ziel untersucht, exaktere Informationen über die Erregungsausbreitung in Abhängigkeit von Membranpotentialänderungen zu gewinnen. In weiteren 12 Versuchen wurde die elektrische Aktivität durch Erhöhung des extrazellulären Widerstandes mit Hilfe einer Gummimembran abgeleitet.

Die erzielten Ergebnisse zeigen, daß im Gegensatz zur Vena portae mit keiner der angewandten Methoden Spikes-Salven von Aortenpräparaten abgeleitet werden konnten. Die an der Rattenaorta abgeleiteten Aktionspotentiale sind vom Schrittmachertyp, mit einer Dauer von 3-10 sec., einer Amplitude von 5-15 mV bei einem durchschnittlichen Membranruhepotential von -35 bis -45 mV.

Mit der angewandten Multielektrodentechnik konnte die Hypotese bestätigt werden, daß jegliche Spannungsänderung von Aortenstreifenpräparaten von der jeweiligen elektrischen Aktivität bestimmt wird.

Different Mechanisms of Activating the Vascular Smooth Muscle by Noradrenaline or Potassium-Depolarization

By U. Peiper, L. Griebel, and W. Wende
Physiologisches Institut der Universität Würzburg, Germany

The contraction of the vascular smooth muscle is induced by releasing calcium ions into the cell. After depolarization of the membrane the calcium-
influx from the extracellular space increases. This mechanism of activating
the contraction is called the electromechanical coupling. By stepwise elevation of the extracellular potassium concentration it is possible to activate
this mechanism gradually (1, 2).

In situ the contraction of the vascular smooth muscle can be induced by releasing noradrenaline from the postganglionic sympathetic nerve fibres. This
contraction, too, is based on the presence of free cytoplasmatic calcium, but
there are at least three ways to explain the release of calcium ions by noradrenaline:
a) Noradrenaline causes a depolarization of the membrane and thus increa-
    ses the calcium-influx (3).
b) Noradrenaline potentiates an existing calcium-influx into the cell without
    an essential change in membrane potential.
    These two supposed mechanisms are based on an enhanced calcium-in-
    flux from the extracellular space after application of noradrenaline.
c) Noradrenaline directly elicites calcium ions from special stores (4). This
    assumption demands two different mechanisms of releasing calcium ions
    by noradrenaline or by depolarization. The electromechanical coupling
    by the calcium ions would be important only to the response to depolari-
    zation, but not for the response to noradrenaline.

A convincing proof of the existence of two different mechanisms presupposes
that it is possible to inhibit the response to depolarization without an essential influence on the response to noradrenaline, and, vice versa, to suppress
the contraction after application of noradrenaline more than the contraction
after depolarization.

In experiments on helical strips of rat aorta the maximum activation of the
preparations was caused by noradrenaline (3 μg/ml) or potassium rich solution (100 mM K) under varied test conditions (fig. 1):

1. When the hydrogen ion concentration of the bath solution was increased,
the noradrenaline-induced tension was far more influenced than the contraction caused by depolarization (5, 6). By blocking the adrenergic α-receptors
(7) or by decrease of the bath temperature (8) the response to noradrenaline
became much smaller, too.
2. As, after application of papaverine, the amplitude of contraction induced
by noradrenaline or depolarization respectively was diminished to the same
degree, papaverine obviously acts on a subsequent section of the activation
mechanism.

3. When the extracellular calcium concentration of the bath solution was changed, the tension development after depolarization, however, varied to a higher degree than after application of noradrenaline (4, 6). When verapamil (iproveratril) was applied in order to inhibit the electromechanical coupling, the influence on the contraction caused by depolarization was more important, too (9).

These results permitted the conclusion that there are two different mechanisms of activating the contraction of the vascular smooth muscle by depolarization or application of noradrenaline. The different dependence on temperature (Q_{10} value concerning the adrenergic mechanism = 2.8; Q_{10} value concerning depolarization = 1.23) indicated the important role of an active process taking part in the activation after application of noradrenaline, while the contraction after depolarization was caused by passive processes on the whole. Only the changes in the contraction amplitude induced by depolarization could be explained by an influence on the electromechanical coupling, and therefore this mechanism of activating the contraction is not important to the activation caused by noradrenaline.

References

1) HUDGINS, P.M., WEISS, G.B.: J. Pharmacol. exp. Therap. 159, 91-97 (1968).
2) WENDE, W., PEIPER, U.: Pflügers Arch. 320, 133-141 (1970).
3) HOLMAN, M.E.: Ergebn. Physiol. 61, 137-179 (1969).
4) HINKE, J.A.M., WILSON, M.C., BURNHAM, S.C.: Amer. J. Physiol. 206, 211-215 (1964).
5) PEIPER, U., WENDE, W.: Pflügers Arch. 314, 14-26 (1970).
6) PEIPER, U., WENDE, W.: Pflügers Arch. 316, R 22 (1970).
7) GRIEBEL, L., PEIPER, U., WENDE, W., GLASER, B., WASNER, E.: Arch. exp. Pathol. Pharmacol. 270, 248-261 (1971).
8) PEIPER, U., GRIEBEL, L., WENDE, W.: Pflügers Arch. 324, 67-78 (1971).
9) PEIPER, U., GRIEBEL, L., WENDE, W.: Pflügers Arch. (in press).

9

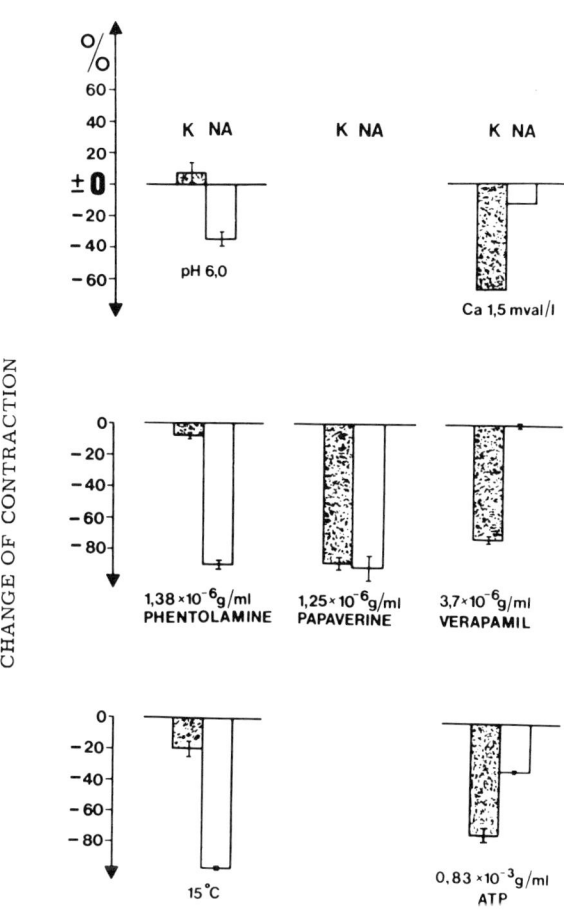

Summarizing review of the different influences of some test conditions on the
two mechanisms of activating the vascular smooth muscle. Left columns: The
contraction induced by noradrenaline (NA) (open columns) is much more affected than the contraction induced by depolarization (filled columns). Middle columns: The influence of papaverine on the contractions induced by noradrenaline or depolarization is almost the same. Right columns: The contraction induced by depolarization is much more affected than the contraction induced by
noradrenaline. The results of a decrease of the calcium concentration are taken from a publication by HINKE et al., 1964.

Central Nervous System Contribution to the Neurogenic Tone of Smooth Muscle

By A. C. Mannard and C. Polosa
Physiology Department, McGill University, Montreal, Canada

Much smooth muscle receives a tonic excitatory drive emanating from the CNS in the form of continual firing of sympathetic preganglionic neurons (SPN). The mechanism of generation, and other features of spike trains emitted by single SPNs during their tonic firing, has been investigated with extracellular microelectrodes in cats. Experiments were conducted on nembutal anaesthetized (either neuraxially intact, NI or high-cervical spinal NS) or unanaesthetized (high-pontine decerebrate, D) cats - thoracotomized, paralysed and artificially ventilated. By the use of time-series analysis (interspike interval histogram, IH and autocorrelogram, ACG) patterns of statistically stationary background firing have been sistematized. These patterns, in conjunction with the reactions of the neurons to antidromic stimuli, were used in hypothesizing stochastic processes responsible for spike initiation.

In D, NI and NS cats 35, 21 and 11 percent, respectively, of SPN had background firing. Roughly half the active cells in both NI and D animals, and all those in NS cats, exhibited the type 1 firing pattern, discharging irregularly, at low rates (0.4-6.0 spikes per sec) and without signs of periodic spike grouping. Generally, the type 1 IH was unimodal with a variable degree of positive skew. All IH showed a range of short interval deficit, RSID (530 msec on average), which also appeared in the ACG as an initial low level, beyond which the contour rose to a uniform level after a phase of "overshoot" and (usually heavily damped) oscillation. Intervals between antidromically evoked firings and the subsequent natural discharge were comparable in duration with the minimum interval measured between action potentials during background activity. Thus, the RSID is not equivalent to the minimum interval between arrivals of powerful input excitatory events (which could dictate SPN discharge times) but reflects a restraint on firing contiguity imposed at the level of the SPN. The time course of post-firing recovery of excitability was measured with antidromic conditioning-testing in electrically silent neurons. It was not uncommon to detect depression lasting longer than a second after a single firing. Since the expected value of RSID is only 530 msec, the relative refractory period is of sufficient duration to be potentially implicated in setting not only the value of RSID, but also the durations of interspike intervals over, perhaps, the entire spectrum. A negative exponential function fits the IH contour only over a range of intervals noticeably longer than the modal interval. The longer interspike waiting times are, therefore, distributed as are the intervals between random events. An interpretation of this is that although, following a spike emission, the probability of firing evolves toward a uniform level, the most common interspike intervals are generated by membrane thre-

Supported by a grant from Canadian Medical Research Council.

shold crossings, which occur while the likelihood of firing is still growing. All experimental observations are deducible from the hypothesis that the timing of background spike initiation is set by a random threshold crossing process operating under the limitation of transient post-impulse refractoriness, with the threshold crossing usually occuring before the recovery process has gone to completion. An effectively random excitatory process could be the result of superposition of spike trains carried in many, independent, converging, presynaptic pathways from spinal and/or supraspinal foci. Individual differences between type 1 patterns in similar preparations could result from differences in input signal intensity and/or SPN recovery properties.

Half the active SPN in D and NI cats had firing patterns with some sort of rhythm (type 2). All IH showed a RSID and many had an initial component resembling a type 1 IH. Cardiac pulse modulation showed up in the single unit spike trains in the associated bimodal or multimodal IH. Respiratory modulation was evident in signals having associated unimodal, bimodal or multimodal IH, all of which shared an ACG period identical to that of the cycle of artificial ventilation. A third form of type 2 firing consisted of well-defined bursts of spikes (recurring, typically, at 5-10 per min). Since the rhythm of burst discharge was not (in several tests) reset by an interpolated antidromic tetanus, the timing of such burst activity was probably dictated by the arrivals of powerful excitation set up, perhaps, by synchronized activation of many presynaptic terminals. Generation of the observed impulse sequences can be interpreted in terms of the operation of a recovery-limited, random process set up by an input with an appropriately modulated intensity, with unimodal, bimodal or multimodal interval distributions emerging, depending upon mean input level, the intensity of modulation and SPN recovery time course. With no control over the input or knowledge of subthreshold activity of the SPN, the hypotheses await corroboration by intracellular recording.

Summary and Discussion

Contraction and dilatation of peripheral vessels are the basis for cardiovascular adjustment. The fundamental mechanisms: Origin and conduction of excitation at the membrane, excitation-contraction coupling, mechanism of contraction are roughly similar to those in the skeletal muscle, but there exist some essential differences. Generally the processes in smooth muscle are much more variable than those in striated muscles. Vascular smooth muscle behaves similar to intestinal smooth muscle which has been investigated more thoroughly for methodological reasons. Studies on vascular smooth muscle are still in the beginning stages. One of the reasons for this is the smallness of the single cells which makes the intracellular recording of membrane potential difficult.

Until now the following facts could be established: The resting transmembranal potential amounts to -30 to -60 mV. Commonly vascular smooth muscle contracts rhythmically. The period duration of the spontaneous contraction lies between 1 second and several minutes up to 1 hour. Fluctuations of the membrane potential are correlated with these contractions (1, 1). Slow waves in the order of seconds can be distinguished from rapid depolarizations in the range of several ms ("spikes"). The spike-potentials induce contractions. In contrast to skeletal muscle contractions without spikes have also been described, sometimes accompanied by long lasting depolarizations. Because of the difficulties of intracellular recording it is still a point of controversy, if during such contractions spike potentials occur which could not be recorded.

The rhythmical excitations originate in changing groups of pace-maker cells. Presumably all vascular smooth muscle cells are able to act as pacemakers. The excitations are propagated by electrical conduction. One or more pacemaker-groups can be active at the same time in one piece of a vessel with more or less synchronization (1, 3).

Some functional differences exist among different blood vessels. Until now the portal vein and the aorta, mostly in the rat, are the best studied objects. Spike-like potential changes could be found only in the portal vein preparation whereas the aorta showed only the slow potential changes. Both can contract rhythmically.

The sympathetic transmitter, nor-adrenaline, acts by way of a depolarization of the membrane, in some cases accompanied by spike potentials, followed by a Ca^{++} influx which induces contraction. But contractions following nor-adrenaline application without changes in membrane potential have also been observed (1, 6). The short action potentials seem to be based on a Ca^{++} inward movement instead of a Na^+ current as in skeletal muscle.

Measurements of transmembranal fluxes of Na^+ and K^+ by the aid of radioactive tracers reveal spontaneous oscillations of ionic exchange, probably caused by rhythmic fluctuations of the activity of ion pumps. These flux oscillations could be the cause of slow potential changes and thus of rhythmic contractions of the smooth muscles (1, 2).

The tone of most vessels is controlled by the sympathetic innervation, media

ted by a continuous but somewhat irregular traffic of impulses in the sympathetic fibres. These impulses originate in preganglionic neurons in a statistically determinable manner under the influence of synaptic inflow from spinal and supraspinal sources with a predominant dependence on baroreceptor
activity (1, 7).

The discussions of the first session were concerned with some of the difficulties in the interpretation of the findings about fundamental processes in vascular smooth muscle membrane physiology and contraction mechanism. Among
others the following questions were discussed:
Do the electrical records of resting and action potentials in vascular smooth
muscle, performed up to now, give the correct picture of these potentials?

It was mentioned that there is a particular difficulty in that even when the
electrical activity is valid recorded from a single cell, the mechanical response must be recorded from the whole tissue.
The answer to the question if there are specialized pacemaker cells in the
wall of the aorta was, that groups of cells rather than a single cell is needed
to generate the activity.

Other questions discussed were:
Are there fundamental differences in the origin of slow and rapid potential
changes? Here it is supposed that the slower potential changes are caused by
Na^+ movements but the rapid ones at least partly by Ca^{++}. In this connection
the unknown value of Ca^{++} equilibrium potential would be interesting.

Is it correct to distinguish tonic contractions from phasic contractions, or are
all tonic contractions tetanus-like states as a result of frequent excitation in
pacemaker cells?

Do there exist several different mechanisms of nor-adrenaline action on the
vascular muscle or is it one fundamental mechanism whose single steps can
be influenced in a different manner by certain agents?

Can the spontaneous rhythmical contractions and/or electrical slow waves be
explained by the measured Na^+ and K^+ flux oscillations or are the latter secondary consequences of other oscillating events e.g. in the membrane permeability?

Generally many fundamental questions still have to be solved before pharmacological and clinical applications are recognizable. On the other hand, the
experimental analysis of these fundamental processes in vascular smooth
muscle physiology undoubtedly will be of great importance for the future understanding of peripheral cardiovascular regulation and thus also for the ability to influence it for the purposes of practical medicine.

Chairmen: W. R. Keatinge, H. P. Koepchen

Role of Cyclic AMP and Ca^{++} in Mechanical and Metabolic Events of Smooth Muscle

By R. Andersson, L. Lundholm, and E. Mohme-Lundholm
Department of Pharmacology, School of Medicine, Linköping, Sweden

In the following investigation we have found that an influence on the c-AMP metabolism is of significance both during relaxation and contraction of bovine mesenteric artery. All tests were done with isometrically contracted muscle. Cyclic AMP was analyzed according to KAKIUCHI and RALL (Mol. Pharmaco. 4, 367-378 (1968)).

The relaxing action of catecholamines which is mediated by a stimulation of adrenergic ß-receptors in vascular and intestinal smooth muscle was preceeded by an increased content of c-AMP (table 1). In intestinal smooth muscle from rabbit there was a clear relation between the increase in c-AMP formation and the degree of relaxation on addition of isoprenaline in different doses. Both the relaxation and the increased c-AMP was blocked by adrenergic ß-blocking agent sotalol.

The same relationship was found in bovine mesenteric artery. Isoprenaline increased the c-AMP content (table 1); this effect was blocked on pretreatment with sotalol. A relaxing action could be produced too by adding c-AMP to the muscle; an effect which have earlier been demonstrated by GEBERT et al. (Ärztl. Forsch. 23, 391-398 (1969)). Both isoprenaline and c-AMP induced other metabolic effects which were associated with the relaxation. The phosphorylase a activity was thus increased and the contents of ATP and CrP decreased markedly.

If Ca^{++} was removed from the suspension solution of the muscle, isoprenaline still increased the c-AMP content (table 1) and the phosphorylase a activity, but had no action on ATP and CrP contents. These studies showed that the relaxation of both vascular and intestinal smooth muscle was associated with a reduction of the ATP content which was dependent on the presence of Ca^{++}.

It is well known that the sarcoplasmic reticulum in skeletal and cardiac muscle accumulate Ca^{++} under utilization of ATP. We have therefore tried to demonstrate the same action on a microsomal fraction from rabbit colon.

We found a Ca^{++}-accumulating protein fraction in intestinal smooth muscle. The accumulation of Ca^{++} by this fraction was significantly stimulated both by isoprenaline and c-AMP. We therefore suggest that the relaxing effect of an adrenergic ß-receptorstimulation is dependent on an increased formation of cyclic AMP and that cyclic AMP stimulates the Ca^{++} accumulation of a microsomal fraction in smooth muscle and thereby reduces the content of free myoplasmic Ca^{++}. When we specifically stimulated adrenergic α-receptors with phenylephrine, we were rather astonished to obtain an increased c-AMP content too in the mesenteric artery. This effect was blocked by dibenamine.

There was, however, a fundamental difference between the effect of ß- and α -receptor stimulation as the α -receptor effect was dependent on the presence of Ca^{++}. In the Ca^{++}-poor preparation both the metabolic effect and contractile action of α -receptor stimulation was blocked, whereas after ß-receptor stimulation the increase of c-AMP content still persisted (table 1).

It is well known that ß-receptor stimulation is dependent on a stimulation of the enzyme adenyl cyclase. We thought that the α -receptor effect on the c-AMP content, was dependent on an inhibition of the c-AMP hydrolyzing enzyme phosphodiesterase. The α -receptor mediated increase of c-AMP was combined with a decrease in the PDE activity (table 1).

To test this hypothesis, we added Ca^{++} and EDTA in varying proportion to a muscle homogenate obtained from a muscle treated with Ca^{++}-free Krebs solution. Free Ca^{++} already in a concentration of 10^{-7} M begun to inhibit PDE. This is about the same concentration of free Ca^{++} which starts to contract the muscle. From other tissues it is known that there exists two different types of PDE. We therefore examined PDE in two concentrations of c-AMP. The PDE activity measured at the lowest concentration (10^{-6}) of c-AMP was most sensible to inhibition of Ca^{++}.

In the following we have tried to summarize our present hypothesis about the relation between Ca^{++}, c-AMP and the mechanical events in smooth muscle.

We suggest that an increased c-AMP will reduce the Ca^{++} in myoplasma of the muscle which leads to relaxation.

Adrenergic α -receptor stimulation will probably release Ca^{++} from this binding sites and contract the muscle as the concentration of free Ca^{++} in the muscle is increased. The PDE activity is thereby inhibited which leads to an increased c-AMP content. This tends to reduce Ca^{++} in the myoplasma. There probably exists a kind of negative feed back between c-AMP and Ca^{++} in smooth muscle.

The well known antagonistic action between α - and ß-receptor stimulation on the muscle tension is explained by this hypothesis too. Whereas α -stimulation releases Ca^{++} in the myoplasma ß-stimulation tends to bind this Ca^{++}.

table 1: Relation between changes of cyclic AMP content, phosphodiesterase activity and mechanical events expressed as per cent of control value, in normal and Ca^{++}-poor muscle-preparations of bovine mesenteric artery and rabbit colon. Significance of the effect is denoted by $*$ = P< 0,05, $**$ = P < 0,01, $***$ = P < 0,001.

Preparation	Adrenergic ß-receptor stim.	Adrenergic α-receptor stim.	Other contrac-ting agent
Mesenteric artery:			
tone	$-19,8 \pm 3,7**$	$+53,2 \pm 7,2**$	$+44,0 \pm 10,0**$
cyclic AMP	$+19,6 \pm 2,1**$	$+25,3 \pm 10,7*$	$+31,6 \pm 12,7*$
cyclic AMP; Ca^{++}-poor			
phosphodiesterase		$-11,1 \pm 2,0**$	$-10,0 \pm 1,8**$
Rabbit colon:			
tone	$-50,3 \pm 3,5**$	$-80,0 \pm 3,0***$	$+11,3 \pm 1,4***$
cyclic AMP	$+46,2 \pm 9,6**$	$-29,3 \pm 11,3*$	$+121,5 \pm 34,2**$
cyclic AMP; Ca^{++}-poor	$+54,3 \pm 10,1**$	$+13,8 \pm 10,7$	$-22,6 \pm 11,3$
phosphodiesterase		$+25,4 \pm 10,6*$	$-14,6 \pm 2,2**$

Cyclic AMP and Vascular Smooth Muscle Function

By L. Triner, Y. Vulliemoz, M. Verosky, D. V. Habif, and G. G. Nahas
College of Physicians and Surgeons of Columbia University, New York, USA

Mediation of the catecholamine relaxing effect in uterine smooth muscle by cyclic AMP was suggested in our previous reports (1, 2). Further experimental work confirmed this and indicated that cyclic AMP has an even broader role in the regulation of smooth muscle function (2): compounds inhibiting cyclic AMP phosphodiesterase and thereby increasing intracellular cyclic AMP, consequently relax the smooth muscle (3); they also potentiate the relaxing effect of compounds which stimulate adenyl cyclase (2). Other compounds, e.g. halothane, acetylcholine, nitroprusside, may also exert part of their functional effects on smooth muscle through the cyclic AMP system. We have shown previously (4) that the cyclic AMP system is present in arterial smooth muscle and is a part of the regulatory mechanism of arterial tone and contractility. The present data add further information about the role of the cyclic AMP system in the control of vascular smooth muscle function.

The formation of cyclic AMP in intact arterial tissue is increased by catecholamines. However, in broken cell preparations of arterial tissue, adenyl cyclase is not activated by catecholamines. This may suggest that the catecholamine-binding sites or their relation to adenyl cyclase in arterial smooth muscle are very fragile. The catecholamine-induced increase of cyclic AMP in intact rat aorta is quantitatively different for each catecholamine; at 5 μ Molar concentration isoproterenol increases cyclic AMP by 150, epinephrine by 90, and norepinephrine by 55 %, respectively (fig. 1).

Similarly, in human gastroepiploic artery, the highest increase in C^{14}-cyclic AMP formation is obtained with isoproterenol and the lowest with norepinephrine, both at 10 μ Molar concentration (2). This stimulatory effect of catecholamines is antagonized by propranolol (10 μ Molar) and enhanced by phentolamine (10 μ Molar). The enhancement by phentolamine is more pronounced with norepinephrine and epinephrine than with isoproterenol and, under these conditions, the C^{14}-cyclic AMP formation is about the same with the three catecholamines.

In order to evaluate the relation of these catecholamine-induced biochemical changes to changes in arterial tone and contractility, catecholamine effects on isolated arterial strip were studied. Both epinephrine and norepinephrine are known to increase tone of the vessel; however, in the presence of an α-adrenergic blocking compound, phentolamine, epinephrine in 0.01 to 0.5 μ Molar concentration decreases the serotonin-induced contraction of the rat aortic strip. This relaxing effect of epinephrine, unmasked by phentolamine, might be related to the enhanced cyclic AMP formation caused by epinephrine in the presence of phentolamine. In contrast (to epinephrine) isoproterenol alone relaxes the artery; in concentrations of 0.01 to 0.5 μ Molar, isoproterenol decreases the serotonin-induced contraction of the rat aortic strip by 15 to 70 %. The relaxing effects of isoproterenol and epinephrine in the presence of phentolamine are quite similar. This seems to be in agreement with the strong isoproterenol-induced stimulatory effect on cyclic AMP formation which is matched by that of epinephrine in the presence of phentolamine. The increase in cyclic AMP brought about by addition of dibutyryl cyclic AMP to the bath is also followed by relaxation of the vessel (2).

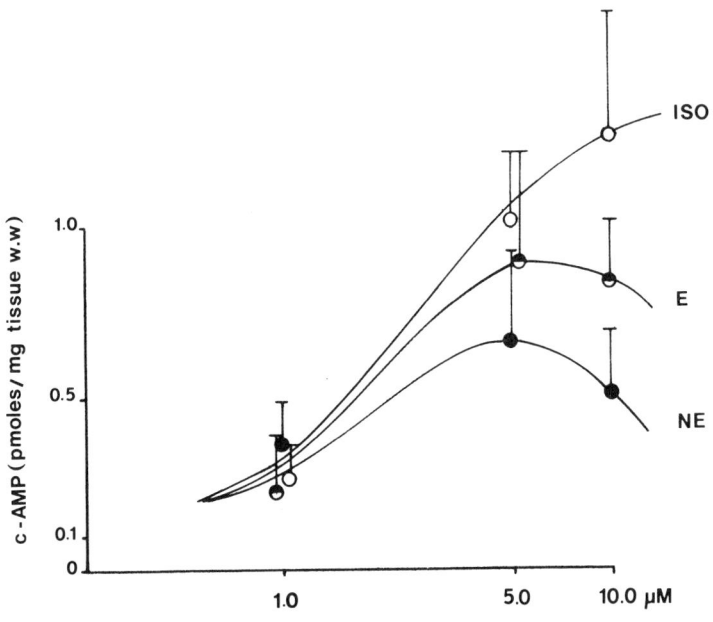

Effect of catecholamines on endogenous cyclic AMP formation in rat aorta. Dose-response curves to isoproterenol O———O, epinephrine ●———●, norepinephrine ●———●. The values represent the net effect (pmols cyclic AMP formed in the presence of theophylline 10^{-2} M/mg tissue wet weight in 10 min minus control values) and are the mean of 9 experiments $\pm$ S.E.

As reported earlier, inhibition of phosphodiesterase, resulting in an increase in intracellular levels of cyclic AMP is accompanied by a decrease in tone and contractility of the vessel (3). The relaxing effect of theophylline and papaverine in rat aortic strip correlates well with the inhibitory effect on phosphodiesterase. When the effect of isoproterenol is tested in aortic strip with the phosphodiesterase inhibited by theophylline or papaverine, a synergistic relaxing effect is observed, as would be expected since both compounds increase the level of cyclic AMP through different mechanisms and, in fact, exert a synergistic rise in cyclic AMP (2).

Since there are differences in the functional effects of vasoactive compounds
in peripheral and central arteries, it was of interest to determine whether
these differences are reflected by different distributions of adenyl cyclase
and phosphodiesterase activity in these vessels. The activity of these enzymes, regulating cyclic AMP level in the cell, was measured along the arterial tree of the dog (2). There is more adenyl cyclase and less phosphodiesterase activity toward the periphery, resulting in a three - to four - fold decrease in phosphodiesterase-adenyl cyclase activity ratio toward the periphery. This is consistent with the fact that peripheral arteries contribute more
to the control of systemic blood pressure and local blood flow than do the di-
stributing arteries. The pattern of enzyme activity distribution may be interpreted in the following sense: The increase in intracellular cyclic AMP resulting from the same activation of adenyl cyclase would be higher in peripheral
vessels, which have a lower phosphodiesterase activity, and thus lead to a
more pronounced dilation of the vessel. Should the amount of adenyl cyclase
be a limiting factor for the relaxing action mediated through this enzyme,
the dilation would again be more pronounced in a peripheral artery. Finally,
the identical inhibitory action on phosphodiesterase could lead to a more marked dilation in a peripheral vessel since there is less activity of the enzyme
hydrolyzing cyclic AMP.

In summary, the enzymes regulating cyclic AMP levels, adenyl cyclase and
phosphodiesterase, are present in various arterial tissues of different species. Catecholamines increase cyclic AMP as a result of their activation of
adenyl cyclase. Activation of adenyl cyclase and/or inhibition of phosphodiesterase resulting in an increase in cyclic AMP is accompanied by vascular
smooth muscle relaxation, indicating that the cyclic AMP system has a role
in the regulation of arterial tone and contractility.

References

1) TRINER, L., OVERWEG, N.I.A., NAHAS, G.G.: Nature 225, 282 (1970).
2) TRINER, L., NAHAS, G.G., VULLIEMOZ, Y., OVERWEG, N.I.A., VE-
    ROSKY, M., HABIF, D.V., NGAI, S.H.: Ann. N.Y. Acad. Sci. 1971
    (in press).
3) TRINER, L., VULLIEMOZ, Y., SCHWARTZ, I., NAHAS, G.G.: Biochem.
    Biophys. Res. Com. 40, 64 (1970).
4) TRINER, L., VULLIEMOZ, Y., VEROSKY, M., HABIF, D.V., NAHAS,
    G.G.: Fed. Proc. 30, 383 (1971).

Die kontraktilen Proteine des Gefäßmuskels und ihre ATPasewirkung

Von L. Laszt

Forschungsinstitut für Cardioangiologie, Freiburg, Schweiz

Man kann den Tonus des quergestreiften Muskels als eine unwillkürliche, te-
tanische Kontraktion auffassen, die wie die willkürliche durch dauernd ein-
laufende Erregungen aufrechterhalten und mit Energieverbrauch verknüpft
ist. Der fundamentale Vorgang bei der Kontraktion ist in beiden Fällen gleich.
Die Gefäßmuskelzellen können stundenlang eine Dauerspannung entwickeln,
die zwei- bis dreimal größer sein kann als bei einem mittleren Blutdruck.
Wir konnten dabei keine Steigerung des O_2-Verbrauches oder der Milchsäure-
bildung nachweisen. Es handelt sich hier wohl um eine Zustandsänderung der
Muskelzellen, wobei Energie nur während der Änderung nicht aber für die
Aufrechterhaltung des Zustandes verbraucht wird. Dies würde darauf hinwei-
sen, daß der Kontraktionsvorgang anders sein muß als beim quergestreiften
Muskel. Morphologisch kann man in der Gefäßwand zwei Muskelzelltypen unterscheiden, einen hellen (chromophoben) und einen dunkeln (chromophilen).
Es ist fraglich, ob es sich hier um zwei Zelltypen mit verschiedener Funktion handelt oder um einen einzelnen in verschiedenem Zustand. Charakteri-
stisch ist nun für den Gefäßmuskel, daß man daraus etwa 80-90 % der kontraktilen Proteine bereits mit Lösungen schwacher Ionenstärke extrahieren
kann und die restlichen, wie diejenigen des quergestreiften Muskels, mit
Lösungen höherer Ionenstärke. Die Myosinkomponente der Actomyosine verschiedener Muskelarten sind physikalisch und chemisch untersucht worden.
Es zeigte sich, daß die physikalischen Eigenschaften bei allen untersuchten
Muskelarten praktisch identisch sind, hingegen zeigen sich große Unterschiede inbezug auf ihre chemischen Eigenschaften. So sind unter anderem Unterschiede in der ATPaseaktivität des Myosins von verschiedenen Muskeln vorhanden, die sich parallel zu der Kontraktionsgeschwindigkeit verhalten. Je
langsamer ein Muskel sich kontrahiert, umso geringer ist die ATPaseaktivität seines Myosins. Wir haben deshalb versucht, zu untersuchen, ob es ge-
lingt, Myosine mit verschiedener ATPaseaktivität aus dem Gefäßmuskel zu
isolieren. Es wurde aus Rindercarotiden zuerst mit einer Lösung schwacher
Ionenstärke das Tonoactomyosin extrahiert und anschließend ein Extrakt mit
Weber-Edsall-Lösung hergestellt. Beide Extrakte wurden in je zwei Teile
geteilt. Während der verschiedenen Dialysen zum Zwecke der Isolierung des
Myosins durch Fraktionierung mit Ammoniumsulfat, wurden zum Schutz der
SH-Gruppen bei je einem Extraktteil jeweils 2 mM Mercaptoaethanol der Di-
alyselösung beigefügt. Wir wollen einfachheitshalber das Myosin aus dem
ersten Extrakt mit Myosin I und dasjenige aus dem zweiten mit Myosin II bezeichnen. In der Abbildung ist graphisch der Ca^{++}-ATPaseaktivitätsverlauf
der verschiedenen Myosinpräparationen in Abhängigkeit von der Ionenstärke

Ich danke dem Schweizerischen Nationalfonds für wissenschaftliche Forschung
für seine Unterstützung.

dargestellt, links der ohne, rechts der mit 2 mM Mercaptoaethanol herge-
stellten (ausgezogene Linie = Myosin I, gestrichelte Linie = Myosin II). Wie
ersichtlich, nimmt die Aktivität des Myosins I, wie bekannt, allmählich zu
und erreicht ein Maximum bei einer Ionenstärke von 0,5 Die Aktivität des
Myosins II ist hingegen praktisch unabhängig von der Ionenstärke. Einen ganz
anderen Aktivitätsverlauf zeigen die Myosine, wenn sie in Gegenwart von Mercaptoaethanol isoliert wurden, nämlich einen sigmoïden. Hervorzuheben ist,
daß bei einer Ionenstärke von 0,1 die ATPaseaktivität des Myosins II etwa
dreimal höher ist, weiterhin, daß die Aktivität des in Gegenwart von Mercaptoaethanol präparierten Myosins I bei einer Erhöhung der Ionenstärke von
0,1 auf 0,5 etwa um das achtfache, diejenige des Myosins II hingegen nur
etwa um das dreifache zunimmt. Das Ultrazentrifugendiagramm zeigt die

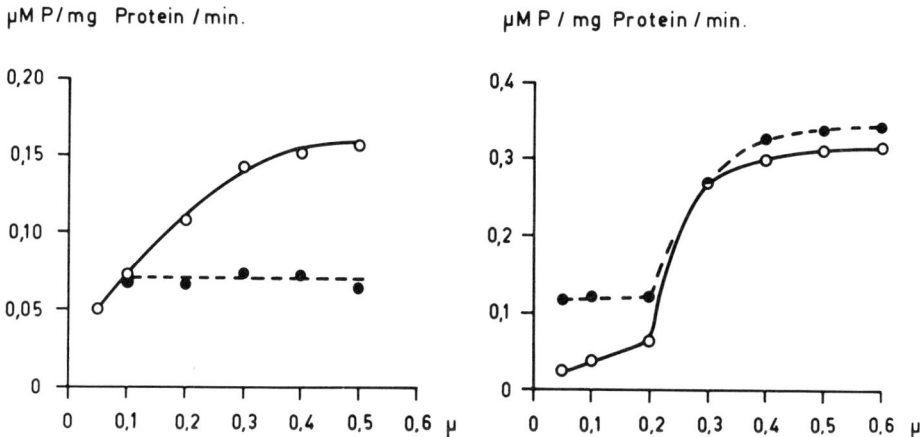

starke Heterogenität des Myosins II. Man kann mindestens vier Hauptgradi-
enten unterscheiden, in Gegenwart von Mercaptoaethanol aber nur zwei. Die
langsam sedimentierende Komponente, die eine Sedimentationskonstante von
3,2 bis 3,5 aufweist, entspricht weder Tropomyosin noch löslichem Kollagen.
Sie könnte dem von RÜEGG und Mitarbeitern gefundenen Extraglobulin entsprechen. Es ist uns aber bisher nicht gelungen, sie abzutrennen. Ob es sich
hier um zwei verschiedene Myosine handelt, oder um dasselbe, das sich in
verschiedenem Zustand befindet, muß weiter untersucht werden. Sicherge-
stellt ist, wie das GASPAR-GODFROID auch schon gefunden hat, daß die Myo-
sin-ATPase des Gefäßmuskels von seinem Aggregatszustand abhängig ist.
Durch Verhinderung der Aggregatbildung durch Mercaptoaethanol erreicht
man beim Gefäßmuskelmyosin die gleichen ATPaseaktivitäten wie beim Myosin des roten quergestreiften Muskels. Meine Arbeitshypothese ist, daß die
tonische Kontraktion auf einem Übergang des Tonoactomyosins vom Sol- in
den Gelzustand zurückzuführen ist, was mit einer Schrumpfung der Zelle ein-
hergeht. Wenn diese Hypothese richtig ist, müsste man mit Mercaptoaethanol eine tonische Kontraktion aufheben können, ohne die phasische zu beeinflussen. Im Versuch in vitro an unbehandelten Ringen aus der Arteria tibialis posterior des Rindes bewirkt die Zugabe von 5 mM Mercaptoaethanol zur
Badeflüssigkeit einen allmählichen Spannungsabfall beziehungsweise eine Va-

sodilatation. 2 mM Aethylmaleinimid, welches im Gegensatz zu Mercaptoaethanol die SH-Gruppen des Myosins oxydiert und die Aggregatbildung fördert, rufen eine langsame Tonuszunahme hervor. Wird die Badeflüssigkeit
der Arteria tibialis posterior von 37 auf 22 $^\circ$C abgekühlt, so tritt zunächst
eine phasische Kontraktion auf, welche allmählich in eine tonische übergeht.
Zusatz von 150 mM Mercaptoaethanol bewirkt die Lösung der Kontraktion. Es
sei betont, daß die phasische Kontraktion nach elektrischer Reizung unbeein-
flußt bleibt. Sollte die tonische Kontraktion auf eine Schrumpfung der Muskelzelle zurückgeführt werden, so müsste die dadurch bedingte Formveränderung
mikroskopisch sichtbar sein. Lichtmikroskopisch sind in erschlafftem Zustand im Längsschnitt keine Unterschiede in Länge und Durchmesser zwischen
hellen und dunkeln Zellen nachweisbar. Während der tonischen Kontraktion
nehmen die dunkeln Zellen sowohl in der Länge als auch in ihrem Durchmesser stark ab, ihr Kern wird dicker. Die hellen Zellen, die durch die Kontraktion mitgezogen werden, werden auch kürzer, aber ihr Durchmesser nimmt
zu. Im elektronenmikroskopischen Bild sehen wir im erschlafften Zustand die
spindelförmigen Zellen nahe aneinander liegend, mit schmalem, langem Kern
Im tonischen Kontraktionszustand sehen wir deutlich Unterschiede zwischen
hellen und dunkeln Zellen. Letztere weisen mehr oder weniger lange Fort-
sätze auf, wie dies schon HAMMERSEN beschrieben hat. An anderen Stellen
ist die Membran ziehharmonikaartig gefältelt, der Extrazellularraum ist dann
vergrößert. In anderen tonisch kontrahierten Zellen wieder sieht man in den
Fortsätzen und in den Membranfalten Vesikel, die wohl Flüssigkeitstropfen
sind. Wir stellen also bei der tonisch kontrahierten Zelle Formveränderun-
gen fest, die sonst bei schrumpfenden Zellen vernehmbar sind.

Effect of Cyclic Nucleosid Phosphates on the Normal and Elevated Myocardial Blood Flow

By E. Roesch, K. Dietmann, W. Juhran, and W. Schaumann
Pharmakologische Laboratorien, Boehringer, Mannheim GmbH, Mannheim, Germany

Cyclic AMP (adenosine 3', 5'-monophosphate) and cyclic GMP (guanosine 3', 5'-monophosphate) were identified in mammalian tissue several years ago and it was suggested that these hormones might play a role in the reactivity of the smooth muscle of the coronary arteries. The present studies were designed to evaluate the effect of these agents and their more lipophilic derivatives on the normal and elevated coronary blood flow (CBF) of conscious dogs. Elevation of CBF was produced by intravenous injection of metrifudil (Th 322). Various amounts of the cyclophosphates up to 80 mg/dog x min were given by direct intracoronary infusion. The normal and pharmacologically enhanced CBF increased further when cyclic AMP and its dibutyryl derivative (cDBA) were given. The effect of cDBA was augmented after theophylline, whereas the effect of cyclic AMP was diminished after theophylline. This suggests that the increase in CBF by cyclic AMP is mainly due to a contamination by non-cyclic adenosine compounds, e.g. 5' AMP which is a very potent coronary dilator. Cyclic GMP was without effect. The possible role of the cyclophosphates in the adaptation of CBF to myocardial metabolism will be discussed.

Monosaccharide Permeability of Arterial Tissue and Smooth Muscle; Effects of Insulin

By H. J. Arnqvist
Department of Pharmacology, School of Medicine, Linköping, Sweden

Patients with diabetes mellitus are more prone to obstructive arterial diseases than other persons. As the development of degenerative arterial diseases probably is associated with changes of the metabolic activity of the vascular

Supported by a grant from Nordisk Insulinfond.

wall (1) it is of great interest to know if the diabetic state influences the metabolic activity of arterial tissue.

In this study (2, 3) the monosaccharide permeability of bovine mesenteric arteries (which contain about 70 % smooth muscle) and rabbit intestinal smooth muscle was investigated. C 14 labelled 3-0-Methylglucose was used to study the membrane transport of monosaccharides. This monosaccharide was not metabolized. Sorbitol C 14 was used to estimate the extracellular space. The distribution (100 x cpm per mg tissue wet weight/cpm per μl incubation medium) of sorbitol reached an equilibrium after 30 min and was confined to about 40-50 % of the wet tissue weight. The initial distribution of 3-0-Methylglucose was the same as that of sorbitol, but reached no equilibrium after 30 min. Instead there was a gradual increase which persisted for at least 150 min. With increasing concentrations of 3-0-Methylglucose in the incubation medium, the tissue distribution decreased when studied after an incubation period of 90 min. The tissue accumulation of C 14 labelled D- and L-glucose were compared. The tissue accumulation of D-glucose-1-C 14 exceeded that of 3-0-Methylglucose while the tissue accumulation of L-glucose-1-C 14 corresponded to that of sorbitol. In both arterial tissue and smooth muscle it was possible to demonstrate counter transport.

These findings show that the membrane permeability of monosaccharides in bovine arterial tissue and rabbit intestinal smooth muscle is characterized by substrate stereospecificity, saturation kinetics and counter transport. This indicates that monosaccharides penetrate the cell membrane of vascular smooth muscle by faciliated diffusion (4).

The tissue concentration of unlabelled glucose was determined by an enzymatic method after an incubation time of 120 min. The glucose concentration in the incubation medium was 11.1 mM. The tissue concentration of glucose in bovine mesenteric arteries and rabbit intestinal smooth muscle corresponded to a distribution of the sorbitol space. It seems therefore probable that free glucose was confined to the extracellular space and that there was no intracellular accumulation of unmetabolized glucose. This indicates that membrane transport has a rate limiting influence on the glucose metabolism in the vascular smooth muscle cell.

The effect of insulin (0.1 U/ml) on the accumulation of C 14 labelled glucose was investigated. In this concentration insulin stimulated significantly ($p < 0.001$) the accumulation of glucose carbon. After an incubation period of 180 min the increase was 14.9 % in bovine mesenteric arteries and 21.7 % in rabbit intestinal smooth muscle. When the membrane transport was studied with 3-0-Methylglucose as a model substance, a small but significant effect ($p < 0.01$) of insulin was evident after 180 min. In skeletal muscle and adipose tissue insulin has a large effect on the membrane transport of monosaccharides. The smooth muscle cell membrane therefore seems rather insensitive to insulin.

These studies indicate that monosaccharides penetrate the vascular smooth muscle cell membrane by faciliated diffusion as in skeletal muscle (5) and adipose tissue (6). There are, however, marked differences. In smooth muscle the glucose penetration of the cell membrane was only to a small degree influenced by insulin. It is therfore probable that the smooth muscle cell metabolism will be more affected by changes in the glucose concentration than by insulin.

References

1) KIRK, J. E.: The Biological Basis of Medicine (E. E. Bittar and N. Bittar, Eds.) Acad. Press 1968, I 493-519.
2) ARNQVIST, H. J.: Acta physiol. scand. (in press).
3) ARNQVIST, H. J.: Acta physiol. scand. (to be published).
4) STEIN, V. D.: The Movements of Molecules Across Cell Membranes. Acad. Press 1967, 126-217.
5) MORGAN, H. E., REGAN, D. M., PARK, C. R.: J. biol. Chem. 239, 369-374 (1964).
6) CROFFORD, O. B., RENOLD, A. E.: J. biol. Chem. 240, 3237-3244 (1965).

Some Effects of the Local Anaesthetic Compound Mepivacaine on Smooth Muscle

By G. Åberg and R. Andersson
Department of Pharmacology, AB Bofors Nobel-Pharma, Mölndal and
Department of Pharmacology, Linköping High School, Linköping, Sweden

Contractile effects of mepivacaine (Carbocaine[R], Bofors Nobel-Pharma) have been demonstrated in vivo in the dental pulp of rats and dogs (POHTO and SCHEININ, 1958, ADLER et al., 1969) and on cutaneous vessels in man (MESNIL de ROCHEMONT and HENSEL, 1960; ÅBERG and ADLER, 1970). Cutaneous resistance vessels are of the single-unit type (BOZLER, 1948) as shown by JOHANSSON and BOHR (1966). In the present investigation the effects of mepivacaine has been studied in vitro in two smooth muscle preparations of the single-unit type.

In the isolated rat portal vein and in the taenia coli of the guinea pig mepivacaine was found to increase the tension and the spike activity but there was no change of the resting membrane potential. After the initial contractile phase that lasted for about 20 min in a Krebs solution with mepivacaine 1 mM, the mechanical activity of the isolated rat portal vein decreased in spite of a persisting and high spike activity.

The L (+)-isomer of mepivacaine was more potent than the D(-)-isomer to increase both the tension and the electrical activity of the muscle. That indicates stereospecific properties of the mechanisms that are influencing those parameters.

The contractile effect of mepivacaine was inhibited by increasing the extracellular calcium concentration. The contraction induced by transferring the muscle into a calcium-poor solution was increased more than 100 per cent by mepivacaine 1 mM. It is therefore probable that an initial event leading to contractions by mepivacaine is that membrane-bound calcium ions are removed and exchanged with the local anaesthetic molecules. That exchange might occur in a similar manner as it has been shown by FEINSTEIN (1964) to occur when local anaesthetic molecules react with phosphate groups of phospholipids in cell-membranes. The increased tension by mepivacaine may thus be of the same kind as in muscles that are transferred to calcium-poor solutions, i.e. by increased migration of sequestered calcium into the cytoplasm when superficially bound calcium is removed (HURWITZ et al., 1967). Mepivacaine increased the content of cyclic 3'5-AMP in calcium-poor preparations. That effect was completely blocked by the adrenergic ß-blocker sotalol. In muscles that were kept in normal Krebs solution and were pretreated with sotalol to block the ß-stimulation, the content of cyclic 3'5-AMP was increased in parallel to the increased tension. The phosphodiesterase activity was decreased in these experiments when the tension and the content of cyclic 3'5-AMP were increased. All these effects were probably due to mepivacaine increasing the "free" calcium concentration in the cytoplasm.

Mepivacaine thus increased the content of cyclic 3'5-AMP both by decreasing the phosphodiesterase activity and by stimulation of adrenergic ß-receptors.

Most of the effects that are reported here on racemic mepivacaine can probably be ascribed to the L(+)-isomer of the compound. The relaxation of potassium-contracted muscles by L-mepivacaine was thus inhibited by sotalol while the relaxations by D-mepivacaine were not influenced by sotalol. The content of cyclic 3'5-AMP was increased in the rat portal vein by the L-isomer of mepivacaine but not by the D-isomer. That indicates different mechanisms for the relaxing effects of the isomers of mepivacaine and we are now using these isomers as tools in our attempts to further analyze the relaxing effects of local anaesthetics.

References

1) ÅBERG, G., ADLER, R.: Svensk tandläk.-T. 63, 671 (1970).
2) ADLER, R., ADLER, G., ÅBERG, G.: Svensk tandläk.-T. 62, 699 (1969).
3) BOZLER, E.: Experientia 4, 213 (1948).
4) FEINSTEIN, M.B.: J. gen. Physiol. 48, 357 (1964).
5) HURWITZ, L., JOINER, P.D., von HAGEN, S.: Amer. J. Physiol. 213, 1299 (1967).
6) JOHANSSON, B., BOHR, D.F.: Amer. J. Physiol. 210, 801 (1966).
7) Du MESNIL de ROCHEMONT, W., HENSEL, H.: Naunyn-Schmiedeberg's Arch. exp. Path. Pharmak. 239, 464 (1960).
8) POHTO, M., SCHEININ, A.: Acta odont. scand. 16, 303 (1958).

Regulation in Vascular Smooth Muscle

By W. E. Russell, W. A. Häcker, and J. C. Rüegg
Institut für Zell-Physiologie der Ruhr-Universität, Bochum, Germany

The recently-reported five-fold activation of a natural actomyosin from vascular smooth muscle by increasing the Ca-ion concentration from 10^{-8}M to 10^{-5}M (SPARROW et al., Amer. J. Physiol. $\underline{219}$, 1366 (1970)) is possible at low ionic strength (~ 0.1) Mg ion concentrations, and is completely absent at < 2 mM. This Ca-sensitivity is similar to that of skeletal muscle in that it depends upon the presence of a regulatory protein system, and in the threshold Ca-concentrations required for activation. This regulatory system from arteries has been found to cross-react with desensitized actomyosin from skeletal muscle, and likewise, the regulatory system of skeletal muscle reacts with that of arterial muscle. Despite its Ca-sensitivity, the natural actomyosin prepared from arteries has been found to contain a slight excess of the inhibitor fraction (troponin B), which can be titrated away. The presence of Mg^{++} and other insolubilizing cations, such as Ca^{++}, has been shown to promote the aggregation of myosin into thick filaments (SCHOENBERG, Tissue Cell $\underline{1}$, 83 (1969)). Vascular natural actomyosin in a medium containing low concentrations of these ions shows the high solubility and absence of Ca-ions sensitivity characteristic of "tonoactomyosin", and a "tonoactomyosin" preparation assayed in the presence of high Mg, shows the characteristic properties of a Ca-ion sensitive natural actomyosin.

Summary and Discussion

Recent investigations have indicated that the elements leading to contraction are of the same nature in smooth muscle as well as in skeletal and cardiac muscle. The contractile proteins in smooth muscle are thus actin and myosin. The myosin component is, however, in some respects different in smooth and skeletal muscle. It has thus a lower ATP-splitting ability, a fact which may explain, that smooth muscle contracts at a slower rate than skeletal muscle. In skeletal muscle, the contraction of actomyosin and its ATPase activity can be triggered by increasing the Ca^{++} concentration up to 10^{-5} M and canceles by EDTA. Under the same conditions, the tonoactomyosin of the vascular smooth muscle shows on the other hand no contraction and no ATPase activity except for the case when 10 mM Mg^{++} are added. Further, in skeletal muscle the contractile and ATP-splitting activity of actomyosin is inhibited by a protein complex of tropomyosin B and troponin, which has been discovered by EBASHI and coworkers. Ca^{++}-ions in a low concentration are able to block the inhibition produced by this protein complex, which explain the contractile actions of Ca^{++}-ions.

RUSSELL, HÄCKER and RÜEGG reported in the discussion on the presence of a similar regulatory protein system in vascular smooth muscle which was inhibited by Ca^{++} too. It is therefore probable that the concentration of Ca^{++} in the myoplasm of smooth muscle regulates the mechanical activity.

Another basic question is the mechanism which regulates the myoplasmic concentration of Ca^{++}. In skeletal muscle the concentration of free Ca^{++}-ions is reduced by a Ca^{++}-accumulating system (the sarcoplasmic reticulum) under utilization of ATP. A Ca^{++}-accumulating microsomal fraction was demonstrated by our group in smooth muscle too. The Ca^{++}-accumulating activity of this fraction was stimulated by adrenergic ß-receptor stimulation and by cyclic AMP. Many other facts reported at this symposium indicated too, that the relaxing and vasodilating effect of adrenergic ß-receptor stimulation was produced by an increased formation of cyclic AMP. Cyclic AMP itself had a relaxing effect. DIETMANN and coworkers reported that dibutyrylcyclic AMP (cDBA) dilated the coronary vessels of the dog. The effect of cDBA was augmented after theophylline, whereas the effect of cyclic AMP was diminished after theophylline. This suggests that the increase in coronary blood flow by cyclic AMP is mainly due to a contamination by non-cyclic adenosine compounds, e.g. 5' AMP which is a very potent coronary dilator. It is well known that the magnitude and sensibility of various vascular beds to adrenergic ß-receptor mediated relaxation is different. The vasodilatation in the skeletal and the coronary vessels is most marked. The findings of VULLIEMOZ and coworkers might explain these differences. They found that the relationship between the activity of the enzyme which convents ATP into cyclic AMP (adenyl cyclase) and the enzyme hydrolyzing cyclic AMP (phosphodiesterase) differed in larger and smaller arteries. The adenyl cyclase activity was most marked in the smaller arteries. In the coronary arteries the adenyl cyclase activity was especially large in relation to the phosphodiesterase activity.

It is probable that many of the clinically used vasodilating agents, as for example papaverine and nitroglycerin, act by increasing the cyclic AMP content of vascular smooth muscle and thereby reduce the free myoplasmic Ca^{++}.

ÅBERG demonstrated that the relaxing action of the local anaesthetic agent mepivacaine relaxed vascular smooth muscle by a mechanism, which at least partly involved an increased formation of cyclic AMP.

Another basal metabolic function of vascular smooth muscle was reported by ARNQVIST, who demonstrated the existence of a monosaccharide transporting mechanism in the muscle membrane. This transporting system was probably rate limiting for the glucose utilization in smooth muscle. It was to a limited degree stimulated by insulin.

LASZT assumed that the tonic contraction of the vascular smooth muscle is caused by transformation of the tonoactomyosin from a sole into the gel. This is connected with shrinking of the cell. Changes in shape and dimension can be visualized microscopically and the delivery of water determined.

Chairmen: L. Laszt, R. Andersson

Use of the Isolated Perfused Arterial Segment for Studying Properties of Vascular Smooth Muscle

By P. Dow and D. L. Davis
Department of Physiology, Medical College of Georgia, Augusta, USA

No preparation is completely objection-free for obtaining results which describe completely and accurately the responses of vascular smooth muscle to stimulation by mechanical forces, nerve stimulation and/or vasoactive hormones. We believe that the isolated perfused segment of a leg artery as we use it offers opportunities for more straightforward and relevant observation and interpretation than many others. The actual hydraulic system in which pressure and flow are being observed can be clearly defined, and the segment is normally tethered and has its normal innervation. Flow can be uncontrolled by the use of existing arterial pressure or kept constant by a proper pump. In the constant flow mode, pressure levels and gradients can be controlled by suitable combinations of clamps and gates. Arterial branches must be avoided; vasa vasorum are a necessary but usually negligible complication. Stimulations are applied either at a small nerve just proximal to the arterial segment or on the distal stump of a crushed sciatic nerve in the presence of succinyl choline. Responses are quantitated in terms of resistance changes of the segment. Under these circumstances we have studied the effects of frequency and duration of stimulus trains, resting intervals, initial stretch, and the interactions of these parameters. We are setting up our equipment for direct photographic recordings to facilitate quantitative estimates of wall tensions.

This research has been supported by grants from the Georgia Heart Association and the USPHS, Grant HE-00240.

Influence of Arterial Smooth Muscle on Visco-Elastic Properties: Measurements in Vitro and on an Electrical Model

By W. J. A. Goedhard

Physiological Laboratory, Free University, Amsterdam, Netherlands

The effects of vascular smooth muscle contraction on visco-elastic properties of the arterial wall are not yet completely understood (WETTERER and
KENNER, 1968).

To study this problem arterial segments, stretched to their natural length,
were examined, before and after administration of norepinephrine. The same
studies were performed with rings, cut from arteries. The tangential Young's
modulus was determined under both static and dynamic conditions. The dynamic modulus was obtained in two ways: By means of analysis of the creep-
phenomenon, i. e. the wall-displacement as a function of time, following a
pressure-step function, and by means of small sinusoidal movements of di-
verse frequencies, superimposed on a steady component.

1. Static Young's Modulus (E_S). On segments of pig's thoracic descending
aorta, E_S was measured by means of pressure steps of 20 mmHg, between
0 and 220 mmHg. When E_S is plotted as a function of pressure, all values
appear to be decreased if smooth muscle is contracted. Since muscle contraction means a small, but obvious decrease of radius, it could be reasoned
that the drop of E_S is a result of this decrease of radius as suggested by
SCHÖNENBERGER and MÜLLER (1960). Examination of the two variables
shows a linear relationship between internal radius and the logarithm of E_S.
It appears that the constructed regression lines (mean of ten experiments)
both for the relaxed and contracted situation, are not the same, but are differing significantly ($P < 0.05$). This means, that the drop of E_S cannot be explained by a decrease of radius only.

Also, the relationship between wall-tension ($T = (p \times R_i)/h$) and the logarithm
of E_S is found to be linear. It appears that the average regression line of
both conditions (relaxed and contracted) run at different levels. Thus, it appears proven that if vascular smooth muscle changes from the relaxed situation into the contracted situation, one is actually dealing with different material porperties.

2. Complex dynamic Young's modulus (E_c). The behaviour of the wall-displacement, following a pressure-step, consists of a rapid rise followed by
a slower exponential-like increase towards a constant level. It is found that
the exponential-like phase can be described as a sum of two exponential po-
wers, with two time constants. The amplitude ratio of rapid and slow phases
is decreased by muscle contraction, i. e. the slow phase is increased in amplitude in comparision to the rapid phase. On the basis of these findings, a
model can be used to determine E_c (WESTERHOF and NOORDERGRAAF,
1970). The model consists of a spring, in parallel to two Maxwell elements
in which the springs represent the elastic, the dashpots the viscous elements.

A Maxwell element consists of a single dashpot and spring in series. The electrical equivalent of the model consists of a capacitor, in series with two RC circuits. By applying sinusoidal voltage-functions (related to pressure) of a range of frequencies and measuring the charge (related to wall-displacement) in the network, E_c, as a function of frequency can be found.

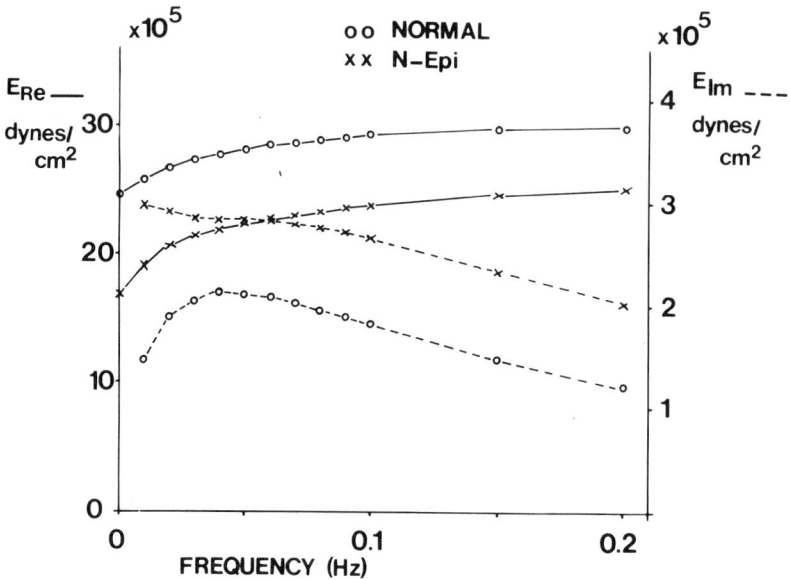

The figure shows E_c, divided into its real (E_{re}) and imaginary (E_{im}) parts. E_{re} is decreased, while E_{im} is increased by muscle contraction, meaning that, within the given frequency range, muscle contraction results in greater energy-losses. On rings of calf's carotid artery, E_c was measured by means of sine-wave functions between 0.017 and 23 Hz. Values of $|E_c|/E_s$, as well as phase difference appear to be increased by contraction of muscle, while it is especially noteworthy, that highest phase-angles are found at the lowest frequencies. By approximating the obtained data of $|E_c|/E_s$, the normalized Young's modulus, in a Bode-plot, it appeared that a good approximation was found by assuming a model, consisting of a capacitor in series with four RC circuits.

Conclusions

a) Contraction of vascular smooth muscle results in a change of "material properties", i.e. a different Young's modulus compared to the relaxed situation. Values of E_s are decreased, whereas values of $|E_c|/E_s$ and phase-difference become increased.

b) Increased values of E_{im} and phase-difference are indicating that contraction of smooth muscle causes higher energy-losses.

c) Linear visco-elasticity can be described by a relatively simple model, consisting of springs and dashpots.

References

1) WETTERER, E., KENNER, Th.: Grundlagen der Dynamik des Arterienpulses. Springer-Verlag, Berlin-Heidelberg-New York, 1968.
2) SCHÖNENBERGER, F., MÜLLER, A.: Über Elastizität und Reaktionsfähigkeit der extracorporalen, im physiologischen Zustand erhaltenen, Rinderaorta. Helv. physiol. pharmacol. Acta 18, 151-173 (1960).
3) WESTERHOF, N., NOORDERGRAAF, A.: Arterial visco-elasticity: A generalized model. J. Biochem. 3, 357-379 (1970).

Die mechanischen Eigenschaften des kontrahierten und nicht kontrahierten Transversalstreifens aus der Kaninchen-Aorta und ihre Analyse anhand eines Analogmodells

Von J. Bücking, M. Herbst, P. Piontek und M. Vonderlage
Physiologisches Institut der Universität Hamburg, Germany

Die bei sinusförmiger Längenänderung von Aortenstreifen gemessenen Kräfte geben Aufschluß über das Dehnungsverhalten der Streifen. Dabei ist es von großem Interesse, das unterschiedliche Verhalten der Streifen im kontrahierten und nicht kontrahierten Zustand zu analysieren. In die bei der Dehnung entwickelte Kraft gehen grundsätzlich drei Komponenten ein: Elastische Kräfte, Reibungskräfte und Beschleunigungskräfte. Das Zusammenwirken dieser drei Komponenten kann in einem mechanischen Netzwerk dargestellt werden, dessen zugehörige Bewegungsgleichungen auf einem Analogrechner gelöst werden können. Da die Beschleunigungskräfte wegen der kleinen Massen gegenüber den hier auftretenden Direktions- und Reibungskräften vernachlässigbar klein sind, wird im folgenden ein Modell verwendet, in dem Beschleunigungskräfte nicht enthalten sind.

Methodik: Transversalstreifen aus der Kaninchenaorta wurden sinusförmige

Längenänderungen unterschiedlicher Frequenz aufgezwungen. Streifenlänge (1): ca 8 mm; Streifenbreite: 1 mm; Größe der Längenänderung: 1,5 mm; Frequenzbereich: $8,2 \times 10^{-4}$ Hz bis 4,17 Hz. Mit Arterenol (10^{-3} g/l) wurden Extrakräfte um 1,5 p induziert, die über mehr als 2 Stunden konstant blieben (kontrahierter Zustand). Zusatz von Papaverin (4×10^{-2} g/l) führte zu einer vollständigen Erschlaffung (nicht kontrahierter Zustand). Weitere Einzelheiten vgl. VONDERLAGE (1970). Das Modell wurde auf einem Analogrechner Telefunken RA 742 gerechnet.

Ergebnisse: 1) Der Elastizitätsmodul des kontrahierten Präparates erhöht sich mit zunehmender Ausgangskraft für alle untersuchten Frequenzen linear, während der des nicht kontrahierten Streifens mit wachsender Ausgangsspannung für alle untersuchten Frequenzen zunehmend ansteigt. 2) Im unteren Frequenzbereich nimmt mit wachsender Frequenz der Elastizitätsmodul des kontrahierten Präparates stark zu, erst zu hohen Frequenzen hin wird er annähernd frequenzunabhängig. Dagegen zeigt der nicht kontrahierte Streifen über den gesamten Frequenzbereich nahezu keine Abhängigkeit von der Frequenz der Längenänderung. 3) Zwischen der Ausgangskraft zu Beginn der Dehnung und dem Minimum der Kraft nach eingeschwungenem Zustand ergibt sich beim kontrahierten Präparat ein Kraftabfall, der mit ansteigender Frequenz zunächst stark und dann nur noch geringfügig zunimmt. 4) Die von Dehnungs- und Entdehnungskurven eingeschlossenen Hysteresisflächen (als Ausdruck der Phasenverschiebung zwischen Länge und Kraft) erreichen bei niedrigen Frequenzen ein Maximum.

Die hier geschilderten Befunde können durch das folgende Modell richtig wiedergegeben werden: Das Modell besteht aus zwei parallel geschalteten Elastizitäten mit den Direktionskräften D_1 und D_2, von denen die eine (D_2) mit einem Dämpfungsglied (Reibungskoeffizient R) in Serie geschaltet ist. D_1, D_2 und R beschreiben das Dehnungsverhalten des nicht kontrahierten Präparates, D_1', D_2' und R' das des kontrahierten Präparates. D_1 und D_1' sind Funktionen der Präparatlänge, die aus experimentell ermittelten Daten gewonnen wurden. D_2' ist bei gegebener Arterenolkonzentration unabhängig von der Präparatlänge. Die Reibungskraft ist geschwindigkeitsproportional. Für das kontrahierte Präparat können wir die in der Tabelle aufgeführte Parameterstellung angeben. Zur Beschreibung des statischen und im wesentlichen auch des dynamischen Dehnungsverhaltens des nicht kontrahierten Muskels genügt D_1, dessen Werte ebenfalls in der Tabelle aufgeführt sind. (Da die Dehnungs- und Entdehnungskurven des nicht kontrahierten Muskels im höheren Frequenzbereich nicht mehr völlig identisch sind, nehmen wir an, daß in diesem Fall ein immer vorhandenes sehr kleines D_2 wirksam wird). Mit diesen Angaben ist das mechanische Verhalten des transversalen Aortenstreifens vollständig beschrieben. Den Übergang vom nicht kontrahierten zum kontrahierten Zustand konnten wir simulieren, indem wir - nach Umschalten auf D_1' - D_2 auf D_2' einstellten. R wurde für alle Zustände konstant gehalten. Es wird damit gezeigt, daß mit zunehmender Muscularis-Kontraktion keine Veränderung des Reibungskoeffizienten erfolgen muß.

Literatur

1) VONDERLAGE, M.: Untersuchungen zur Veränderung des Elastizitätsmoduls der Gefäßwand mit zunehmender Muscularis-Kontraktion. 1970 Habil Schrift, Universität Hamburg.

i	0 - 7	8	9	10	11	12	13	14	15	16	17	dim
D_1	0	0	0	0	3,1	4,7	5,6	6,8	8,9	12,1	16,4	10^3 dyn/cm
D_1'	0	3,6	5,5	6,7	7,4	7,7	8,2	9,2	10,7	12,8	16,3	10^3 dyn/cm
D_2						0						10^3 dyn/cm
D_2'						6						10^3 dyn/cm
R						16						10^3 dyn/cm
R'						16						10^3 dyn · s/cm

$$l = i \cdot \Delta l \qquad \Delta l = 0,077 \text{ cm}$$

Effects of Smooth Muscle Contraction on Diameter and Elasticity of Ascending Aorta

By H. Aars

Institute of Experimental Medical Research, University of Oslo, Ulleval Hospital, Oslo, Norway

The effect of smooth muscle contraction on diameter and elasticity of the ascending aorta was examined in 18 anesthetized rabbits (3). Aortic diameter was measured based on the transit time of ultrasonic pulses between two crystals fastened to aorta 3-15 days before (1), and blood pressure was recorded from a catheter in the right common carotid artery. Control observations were obtained at various blood pressure levels by producing stepwise changes of the blood volume. At diastolic pressures from 40 to 90 mmHg i.v. infusion of noradrenaline (4-6 µg/kg · min) caused a 6.6 % average reduction in diastolic aortic diameter. In an earlier study (2), the smaller doses of 1-2 µg/kg · min were found to produce a mean reduction of 2.5 %. The effect of the smaller doses disappeared completely if the study was performed during an acute thoracotomy, illustrating that the operative trauma and exposure of the vessel had made aorta unresponsive to noradrenaline added throug infusion. To ensure a proper control state, aortic diameter should therefore be measured within the intact, undisturbed thorax.

The influence of noradrenaline of the elasticity of aorta was assessed by the use of the pressure-strain elastic modulus, E_p · ($E_p = (\Delta P)/(\Delta D/D)$ dyn/cm E_p in the control state ranged from about $0.5 \cdot 10^6$ (at 40 mmHg) to $2 \cdot 10^6$ dyn/cm^2 (at 90 mmHg). E_p was significantly increased by noradrenaline at 40-60 mmHg, up to 67 %, but showed no changes at 70-90 mmHg. Similar effects were found on the pulse wave velocity (PWV), which could be calculated from E_p. Control values of PWV ranged from 5 (at 40 mmHg) to 10 m/sec (at 90 mmHg), and both E_p and PWV were thus higher than commonly found in man and in the dog. This might partly be due to the much higher heart rate in the rabbit.

It is concluded that, at comparable pressures, contraction of aortic smooth muscle tissue, induced by i.v. infusion of 4-6 µg/kg · min of noradrenaline, in the rabbit produces a significant constriction of the ascending aorta at diastolic pressures ranging from 40 to 90 mmHg, and, at low pressures only, increases dynamic stiffness of the vessel.

References

1) AARS, H.: Relationship between blood pressure and diameter of ascending aorta in normal and hypertensive rabbits. Acta physiol. scand. 75, 397-405 (1969).
2) AARS, H.: Effects of altered smooth muscle tone on aortic diameter and aortic baroreceptor activity in anesthetized rabbits. Circulat. Res. 28, 254-262 (1971).

3) AARS, H.: Diameter and elasticity of the ascending aorta during infusion of noradrenaline. Acta physiol. scand. (in press).

Correlation between Distensibility Characteristics of Individual Mesenteric Microvessels and the Pressure-Flow-Relationship of the Mesenteric Vascular Bed

By P. Gaehtgens and U. Uekermann
Institut für normale und pathologische Physiologie, Köln, Germany

Isolated segments of canine mesenteric membrane were perfused at constant pressures and the pressure-flow-relationship determined between 170 mmHg and 0 mmHg. The pressure-flow-relationships obtained in 29 experiments were convex to the pressure axis. Blood flow at 100 mmHg averaged 0.39 ± 0.1 ml/min · g dry weight; this value is comparable with data obtained for mesenteric blood flow in situ (1, 2). In addition, the dimensional changes of arteriolar and venular microvessels (12-148 μ) in the same tissue specimens were determined microphotographically:
1) Arterial microvessels showed significant changes of diameter (approx. 15 % over the pressure range studied); approximately 50 % of these vessels exhibited distensibility characteristics which were characterized by a plateau of virtually unchanged diameter between 70 mmHg and 170 mmHg. Changes of microvascular length (up to 25 % of maximum length) were seen only in those arterioles which were not oriented parallel to the prevailing tension in the mesenteric membrane. The moduli of volume elasticity calculated from these data were found to average $4 \cdot 10^6$ dynes/cm^2 at pressures of 100 mmHg.
2) Venous microvessels were found to show changes of length which were smaller (6 % over the pressure range) and correlated to the length changes of the accompanying arterioles. Venular diameter changed by 25 % over the range of arterial pressures studied. From measurements of venular distensibility upon changes of venous pressure moduli of volume elasticity were calculated which averaged $0.7 \cdot 10^5$ dynes/cm^2 at venous pressures of 10-20 mmHg. Venular distensibility was limited at pressures exceeding 30 mm Hg.

The comparison of the pressure-resistance-relation of the entire bed with that calculated from individual microvessels (fig. 1) showed that the arterioles studied represent the stiffest elements of the mesenteric microvasculature (capillaries excluded). The discrepancy of the two pressure-resistance curves is only to an insignificant degree reduced by taking into account the

contribution of the smallest venules to total resistance. It is concluded that
a) larger arterial and/or venous vessels contribute significantly to the pressure-induced changes of total resistance of the vascular system,
b) the distensibility of terminal mesenteric microvessels is to a large degree
influenced by the direction and amount of tension in the surrounding extravascular tissue structures.

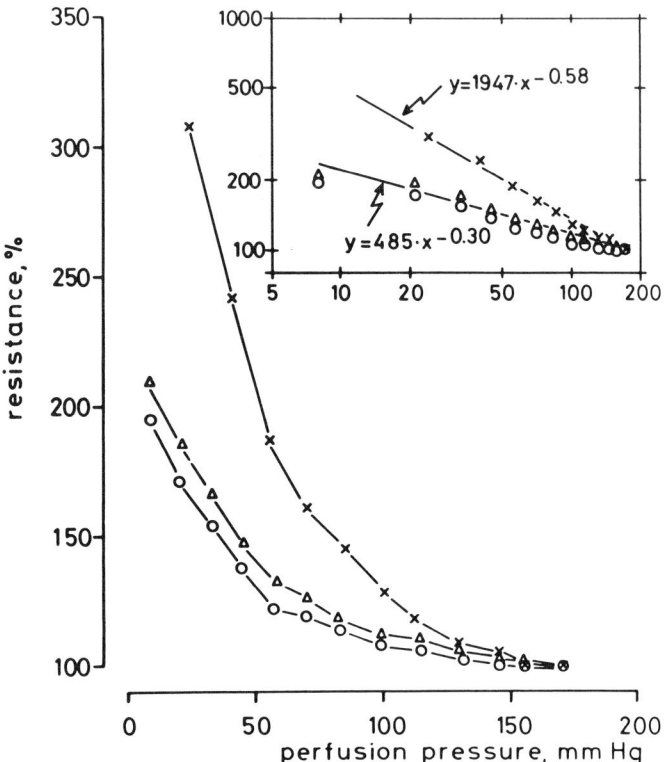

Pressure-resistance-relation calculated from measurements of total flow (x)
and distensibility characteristics of individual arterial microvessels alone (o)
and in combination with the distensibility characteristics of venous microvessels (Δ) in the isolated and perfused mesenteric membrane. The insert shows
a log-log-plot of these relations; the difference between the linear regression
fitted to these data is statistically significant (p< 0.001).

References

1) GRIM, E.: The flow of blood in the mesenteric vessels. In: Handbook of
    Physiology. W. F. Hamilton and P. Dow (Eds.) Washington D.C. Ameri
    can Physiological Society, Section 2. Circulation 2, 1439-1456 (1963).
2) RAYNER, R.L., MacLEAN, L.D., GRIM, E.: Intestinal tissue blood
    flow in shock due to endotoxin. Circulat. Res. 8, 1212-1217 (1960).

The Behavior of the Isolated Rat Tail Artery towards Sinusoidal Strains Investigated at Different Vascular Smooth Muscle Tones

By F. O. Grube, R. D. Bauer, and Th. Pasch

II. Physiologisches Institut der Universität Erlangen-Nürnberg, Germany

For more than 30 years, the extent to which the vascular smooth muscle tone may influence the elastic modulus of the vessel wall has been under discussion (7). As the modulus is dependent on the circumferential stress $_t$ (1, 2), a comparison between values of moduli is only meaningful if the actual $_t$ is considered. To elucidate the problem, we examined the rat tail artery whose wall contains about 75 % of smooth muscle (4). A segment free of sidebranches was excised, cannulated at both ends, stretched to the in-situ length of about 6 mm, and put in Tyrode's solution (37^0 C) aerated with carbogen (95 % O_2, 5 % CO_2). The interior of the arterial segment was connected at one end to a chamber where rhythmical volume changes were produced. The other end was connected to a manometer (Statham P 23 Gb). Chamber, artery, and manometer were filled with Tyrode's solution free of air bubbles. The frequency of the volume changes was varied from 0.018 to 25 Hz. At first we added norepinephrine ($3-4 \cdot 10^{-3}$ g/1) to the bath afterwards replaced by papaverine (10^{-4} g/1). The absolute value E_t of the circumferential elastic modulus was calculated from the recorded pressure volume relations. Using the differential definition of FRANK (6), we calculated E_t of the middle layer of thick-walled vessels, held in longitudinal constraint, by the formula:

$$E_t = \frac{p \cdot r_i}{h} \cdot \frac{r_m}{\Delta r_m} \cdot \left(\frac{\Delta p}{p} + \frac{\Delta r_i}{r_i} \right) \cdot (1 - \mu^2),$$

where p = transmural pressure; Δ p = amplitude of sinusoidal changes in transmural pressure; $r_m = \sqrt{r_a \cdot r_i}$ = radius of the middle layer of the wall; r_a = outer radius; r_i = internal radius; Δr_m, Δr_i = amplitude of sinusoidal changes in r_m and r_i, respectively; h = wall thickness; μ = Poisson's ratio (for simplicity's sake, assumed to be 0.5). The formula does not take into account that E_t is also dependent on the radial elastic modulus E_r. If $E_r \geq E_t$, this error amounts to less than 10 %. The circumferential elastic modulus is given by a complex number composed of a real part, the so-called dynamic elastic modulus $E_d = E_t \cdot \cos \varphi$, and an imaginary part, the so-called loss modulus $\omega \eta_w = E_t \cdot \sin \varphi$ $\cdot \varphi$ is the phase angle by which Δ p leads Δ V where Δ V = amplitude of sinusoidal volume changes. ω is the angular frequency and η_w the coefficient of wall viscosity.

The results of these experiments: At the same pressure and the same frequency, E_t is far smaller after treatment with norepinephrine than it is after the application of papaverine, e.g. for norepinephrine $5 \cdot 10^5$ dynes/cm^2 and for papaverine $17 \cdot 10^5$ dynes/cm^2 if p = 100 mmHg and f = 2 Hz. To clarify the question as to what extent this increase of modulus is due to an enlargement of σ_t, induced by an increase of the diameter, we related E_d and $\omega\eta_w$ to σ_t, forming the dimensionless quotients E_d/σ_t and $\omega\eta_w/\sigma_t$ and regarding these quotients as functions of frequency. The calculation of these quotients is meaningful because there is a linear relation between E_d and σ_t as well as between $\omega\eta_w$ and σ_t. The result is shown in fig. 1. When the vascular smooth muscle was strongly constricted, E_d/σ_t was found to rise slightly with frequency and to be evidently smaller than under the influence of papaverine where the frequency dependence of E_d/σ_t is more clearly to be seen. Assuming that the elastic modulus of the wall is essentially determined by the smooth muscle in case of marked constriction, it must be concluded that the elastic modulus of the active vascular smooth muscle is smaller than the one of the elastic components of the wall. This results stands in accordance with (5). The expression $\omega\eta_w/\sigma_t$ shows, in principle, the well-known frequency dependence (3) and proves to be independent of the smooth muscle tone. In other words, if the vascular smooth muscle is excited, the viscous part of the circumferential elastic modulus, related to the actual value of E_t, is greater than under the condition of relaxed vascular smooth muscle.

References

1) FRANK, O.: Z. Biol. 71, 255-272 (1920).
2) FRANK, O.: S.-B. Ges. Morph. Physiol. München 37, 23-32 (1926).
3) HARDUNG, V.: Helv. physiol. pharmacol. Acta 11, 194-211 (1953).
4) HINKE, J.A.M., WILSON, M.L.: Amer. J. Physiol. 203, 1153-1160 (1962).
5) LASZT, L.: Angiologia 5, 14-27 (1968).
6) WETTERER, E., KENNER, T.: Grundlagen der Dynamik des Arterien-pulses. Springer-Verlag:Berlin-Heidelberg-New York, 1968.
7) WEZLER, K., BÖGER, A.: Ergebn. Physiol. 41, 291-606 (1939).

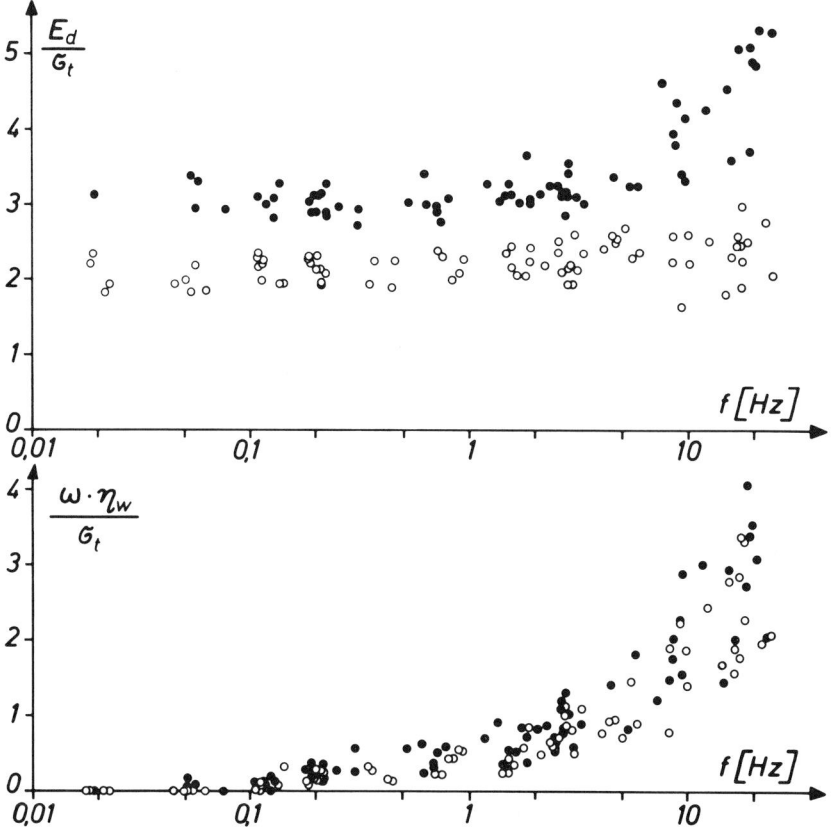

Dynamic elastic behavior of the wall of the rat tail artery after treatment
with papaverine (●) and norepinephrine (o). Top: quotient of dynamic elastic
modulus and circumferential stress (E_d/σ_t). Bottom: quotient of loss modulus and circumferential stress ($\omega\eta_w/\sigma_t$). Abscissa: frequency f.
Measurements from experiments on three different animals

Active/Passive Factors in Local Blood Flow Regulation

By S. Rodbard
Department of Cardiology, Duarte, California, USA

Our recent studies have explored the possibility that blood flow may be regulated locally by mechanical factors operating at the capillary. Our evidence has indicated that flow varies with the transmural capillary pressure in heart, skeletal muscle, kidney and lung. In models, this can be shown to depend on flexible, permeable blood capillaries enclosed with extracapillary fluid and parenchymal cells in a compliant capsule. Such systems exhibit critical closing, basal tone, the hyperemias, autoregulations and other phenomena of the peripheral circulation, without need to introduce active arteriolar smooth muscle contraction. Related studies have shown that smooth muscle constriction produces remarkably large, complex effects on the lumen cross sectional area and thus even more complex effect on vascular conductance ($= \text{resistance}^{-1}$). This suggests that each arteriole probably can not regulate flow without the aid of a very complex local computer. It is suggested that each arteriole is either fully opened or closed. Vascular conductance may then indicate the percentage of available arterioles open at any instant. Vasoconstrictor agents would be viewed as decreasing the percentage of vessels that are open. Intermittence of closure for any single vessel would prevent prolonged ischemia of a given vascular bed. This approach could account for changes in conductance as adequately as the presently accepted belief that such vessels are capable of grading the magnitude of constriction. Related studies of such arteriolar closure on capillary conductance indicate that such an arrangement will improve the exchange of extracapillary fluids and the wash-in and wash-out of metabolic materials.

Aided by HE 08721.

Summary and Discussion

Continued interest in vascular elasticity indicates that the concepts of Otto
Frank as to the importance of this property have not been forgotten. Elasticity may be described in tension/length or pressure/volume units, or expressed as a more universal modulus. The former are direct reportings of the
measurements without manipulation, while the latter have the value of per-
mitting comparison of different experiments and different tissues even though
their validity may be compromised by uncertainty as to the exact geometry
of the measured segment. The inferences that one might draw from volume
measurements can be significantly modified when the data are converted to
tangential wall tension.

In vascular tissue in which the muscular element is relaxed, investigators
agree on a relatively simple though non-linear elastic behavior with convexity toward the length or volume axis. The functionally significant questions,
however, relate to the influence of vascular tone, which is certainly not the
exclusive property of the arteriolar "resistance" vessels but is readily de-
monstrable in "elastic" vessels such as the aorta (WEZLER, 1937, AARS)
and muscular arteries (WEZLER - SCHLÜTER, 1953, DOW, GRUBE et al.).
Active contraction of the vascular smooth muscle introduces a very marked
viscous component, most conspicuously evidenced by the development of wide
hysteresis loops when the tissue is subjected to cyclical stretching (AARS).
With small stretches in the pressure ranges most commonly studied, muscular contraction produces a significant change in stiffness (modulus). The quan-
titation of this change in stiffness, however, is highly dependent upon the pattern and rate of stretch, and with some techniques an actual decrease in the
modulus may be observed with muscle contraction (WEZLER - BÖGER, 1939,
GRUBE et al.). Vascular physiologists should not forget the elementary facts
they learned about frog muscle: Active contraction significantly shortens the
length of a lightly loaded muscle, but at maximal stretch the active and passive tension curves converge. It is thus to be anticipated that dynamic stiffness as measured by small stretches should be increased by muscle contraction, even though over the full range of tension there must be a greater change
in length in the contracted tissue. As a first approximation, it is attractive
to resolve this problem in terms of visco-elastic models (BUCKING et al.) or
their electrical equivalents (GOEDHARD) in which the frequency characteristics of the wall can be rigorously defined. The adequacy of these models is
open to question because the tissue shows a "memory" for its previous history
of stretching which exceeds simple visco-elastic theory (WEZLER, 1937).
The responsiveness of the contractile element also seems to be influenced by
the recent history of the tissue (DOW).

There is a danger that we may become so entranced with the challenge of de-
fining the physical properties of the material of the wall as to forget the physiologically important questions. For example, isolated blood vessels have
lost the tethering action of supporting tissue which may be of considerable
functional importance (GAEHTGENS and UEKERMANN). Also, in most studies of vascular elasticity there is a notable absence of evidence of the autoregulatory "Bayliss effect". While one cannot decry the value of clarifying
the mechanical properties of the blood vessel wall, sophisticated Bode plots
of the mechanical properties of the tissue may have limited relevance to the

behavior of blood vessels in the body if their major response to stretch is an active contraction of the smooth muscle (WEZLER- SINN and WEZLER - SCHLÜTER, 1953). This caution is particularly important if vascular behavior is not only dependent upon wall properties, but also upon a complex of transmural dynamics influenced by changes in tissue pressure (RODBARD).

Chairmen: K. Wezler, R. S. Alexander

Factors Determining Movement and Distribution of the Transmitter in the Vascular Wall

By J. Török and J. A. Bevan

Institute of Normal and Pathological Physiology, Slovak Academy of Sciences, Bratislava, Czechoslovakia, and Department of Pharmacology, UCLA School of Medicine, Los Angeles, California, USA

The entry and movement of norepinephrine through the vessel wall and the early time course and distribution of both uptake processes of norepinephrine, neuronal and extraneuronal, through the thickness of rabbit thoracic aorta has been investigated using an isotopic frozen section technic (1). Studies were performed on helical strips incubated in Krebs bicarbonate solution containing ^{3}H-norepinephrine (^{3}H-NE) $0.75 \cdot 10^{-6}$ M for varying periods of time up to 10 min, which period was in excess of the time taken to reach the plateau of contraction to NE.

The patterns entry of ^{3}H-NE through the intima and the adventitia are different. The slower movement and lower accumulation of ^{3}H-NE in the tunica media after intimal exposure is probably the consequence of structural constituents of vessel wall, such as the elastic lamellae (2). The more rapid entry through the adventitia is consistent with existence of functionally large communicating spaces of channels in this tunica which rapidly fill with ^{3}H-NE (1).

The total (intra- and extraneuronal) uptake of ^{3}H-material increased with incubation time. The transmitter accumulation after short exposure time preferably occupies the extracellular space in the vessel and can be easily completely washed out. Detectable intraneuronal uptake could be observed as soon as 1 min after exposure in distribution studies and at that time the concentration of ^{3}H-NE in the region of the adventitio-medial junction was 0.03 μg/g. Extraneuronal uptake was detectable not earlier than after 5 min of exposure and was found to increase with incubation time. It occured mainly in the tunica media.

No difference in accumulation of ^{3}H-material and its distribution through the thickness of vascular wall was observed after l- or dl-^{3}H-norepinephrine up to 10 min of incubation.

References

1) TÖRÖK, J., BEVAN, J.A.: J. Pharmacol. exp. Ther. 177, 613-620 (1971).
2) BEVAN, J.A., TÖRÖK, J.: Circulat. Res. 27, 325-331 (1970).

Regional Variation in Adrenergic Neuroeffector Mechanism in Blood Vessels

By J. A. Bevan and C. Su
Department of Pharmacology, UCLA School of Medicine, Los Angeles,
California, USA

The adrenergic neuroeffector mechanism in different macroscopic vessels
has been examined, and some of the factors that contribute to its variation
amongst different blood vessels have been defined. In general, vessels from
the rabbit were studied by in vitro techniques.

Differing density and distribution of adrenergic innervation within the walls
of various vessels, and the variation in the structure of these walls have been
noted. Some of the consequences of this variation on transmitter mechanisms,
in particular, transmitter release and uptake and also the proportion of the
released transmitter to enter the tunica media have been explored. This latter component is dependent amongst other things on the relative resistance
to diffusion in the adventitia and media. The effective concentration of the
transmitter once it has entered the media is influenced by the relative affinity and capacity of the various tissue binding sites and the activity of the various metabolic processes.

The nature of the vascular muscle response to the adrenergic transmitter is
not uniform. Some vessels respond to NE by a slow, tonically maintained
contraction; others, particularly those that show spontaneous rhythmic activity, by changes primarily in rate and amplitude. Other vessels show inter-
mediate effects in which both tonic and phasic responses are seen.

Studies of the responses of a variety of vessels to l-norepinephrine and a series of directly-acting chemical analogues show that neither their sensitivity
to l-norepinephrine, nor the relative potency of the analogues is constant
amongst different vessels. This suggests that in different regions, the sensitivity of vascular muscle to the transmitter and the specificity of the adrenergic receptor itself is not constant.

This work was supported by grants from the USPHS., AMAERF, and LACHA.

Regulation der Gefäßmuskelkontraktion durch zwei in der Gefäßwand vorkommende vasoaktive Stoffe

Von L. Laszt

Forschungsinstitut für Cardioangiologie, Freiburg, Schweiz

Die kontraktilen Eigenschaften der Muskelzellen verschiedener Gefäßbereiche sind wie bekannt unterschiedlich. In einem und demselben Gefäßbereich nimmt nach übermaximaler elektrischer Reizung peripherwärts die Latenzzeit ab. Gleichzeitig nehmen die Kontraktionsgeschwindigkeit und die Muskelkraft zu. Es besteht keine Parallelität zwischen Kontraktionsgeschwindigkeit und ATPaseaktivität des entsprechenden Myosins, wie dies bei anderen Muskeln der Fall ist. Diese Eigenart des kontraktilen Verhaltens der Gefäßmuskelzelle führen wir auf ihren Gehalt an zwei Vermittlerstoffen, die während der Kontraktion freigesetzt werden, zurück. Einer davon verkürzt die Latenzzeit und erhöht Kontraktionsgeschwindigkeit und Muskelkraft, der andere hemmt die Kontraktion und fördert die Erschlaffung. Bei Versuchen in vitro tritt ein Teil dieser Stoffe während der Kontraktion in die Badeflüssigkeit über, wo sie während eines Arbeitszyklus am Meerschweinchenileum nachgewiesen werden können. Der kontraktionsfördernde Stoff ruft nämlich am Meerschweinchenileum eine Kontraktion hervor, indessen der erschlaffungsfördernde Stoff antagonistisch wirkt.

Zum Nachweis der Freisetzung beider Stoffe während eines Arbeitszyklus haben wir zwei Methoden angewandt: Bei der einen verwenden wir eine Doppelkammer. In der einen Kammer befindet sich, zwischen zwei Stiften fixiert, ein aus der Arteria tibialis posterior des Rindes entnommener Ring, der elektrisch gereizt wird. Die beiden Stifte dienen gleichzeitig als Reizelektrode. In der zweiten Kammer ist ein Stück Meerschweinchenileum angebracht. Zur Registrierung werden Arterie und Dünndarm mit je einem Transducer verbunden und die Anodenstromschwankungen mit Hilfe eines SIEMENS-OSCILLOFIL aufgenommen. Während des Versuchs wird die Badeflüssigkeit (50 ml) mit einer Zentrifugalpumpe zwischen den beiden Kammern in Zirkulation gebracht. Der Nachteil dieser Methode besteht darin, daß die Zirkulation nicht genügend rasch ist, sodaß die zeitlichen Verhältnisse während der Kontraktion nicht getreu wiedergegeben werden. Bei der zweiten Methode - links dargestellt - befinden sich sowohl der Arterienring als auch der Dünndarm in derselben Kammer (Badeflüssigkeit 100 ml). Wird bei der Doppelkammermethode die Arterie gereizt, so beginnt sich die Darmwand, wenige Sekunden nachdem die Arterie maximal kontrahiert ist, auch zu kontrahieren. Die Verkürzung nimmt zuerst rasch, dann allmählich zu. Nach Beginn der Erschlaffung des Arterienringes tritt auch diejenige des Dünndarms ein, wobei letzterer regelmäßig zuerst unter seiner Ausgangslänge bleibt und erst allmählich wieder zu dieser zurückkehrt. Anders ist der Kontraktionsverlauf

Ich danke dem Schweizerischen Nationalfonds für wissenschaftliche Forschung für seine Unterstützung.

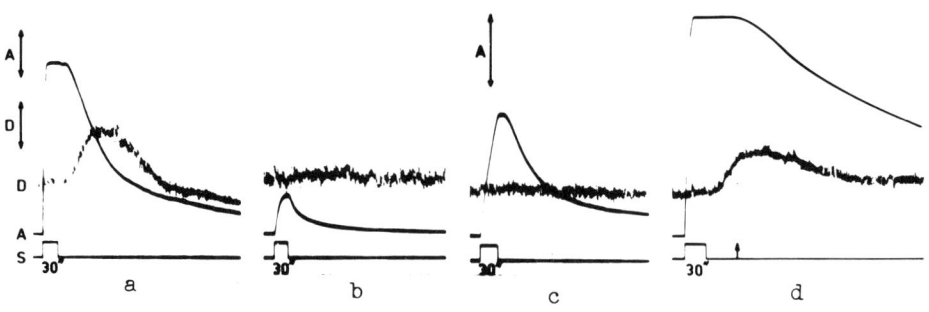

des Dünndarms bei Verwendung der Einzelkammermethode. Wie Abb. (a)
zeigt, beginnt die Kontraktion in diesem Fall auch nachdem die Arterie sich
maximal kontrahiert hat. Sie verläuft aber steiler. Nach Erreichung eines
Gipfels erschlafft der Dünndarm auch hier wieder zunächst unter seine Ausgangslänge. Obwohl wir zur Erregung eine Trägerfrequenz von 1000 Hz be-
nützten, die durch Unterbrechungen in der Zahl von 30/sec moduliert wurde,
stellt sich aber bei Anwendung des Einkammersystems die Frage, ob die
Kontraktion des Dünndarms tatsächlich durch Freisetzung von Stoffen aus
dem Gefäßring oder möglicherweise durch Stromschleifen hervorgerufen wird,
weshalb wir eine Reihe von Kontrollversuchen durchgeführt haben. Wenn wir
nun in der Kammer nur den Dünndarm und kein Gefäß anbringen und dann 30
Sekunden den Strom schließen, so tritt keine Kontraktion des Dünndarms ein.
Legt man einen Gefäßring in Kalium-Tyrodelösung, so wird der größte Teil
der Stoffe in die Tyrodelösung abgegeben. Lassen wir jetzt denselben Gefäßring wieder in Natrium-Tyrodelösung erschlaffen und fixieren ihn dann im
Apparat, so sehen wir nach elektrischer Reizung, wie es die Abb. (b) zeigt,
daß die Kontraktion des Gefäßringes viel geringer ist und daß der Dünndarm
nicht beeinflußt wird. Bei anderen Versuchen haben wir den Gefäßring, nachdem der Dünndarm begonnen hat, sich zu kontrahieren, aus der Badeflüssigkeit herausgenommen. In diesem Fall bleibt der Dünndarm kontrahiert und
kommt nicht auf seine Ausgangslänge zurück (d). Weiterhin haben wir gefunden, daß Pferdegefäße die genannten Stoffe nicht enthalten. Reizen wir unter
gleichen Bedingungen einen aus der Arteria tibialis des Pferdes entnommen-
en Ring, so tritt keine Kontraktion des Dünndarms ein (c).

Zugabe von 10^{-8} Pyribenzamin zur Badeflüssigkeit, welches die Histamin-
wirkung vollständig aufhebt, hat keinen Einfluß auf die Dünndarmkontraktion.
Hingegen tritt, wenn wir 3 γ/ml Regitin zusetzen, eine rapide Lösung der
Kontraktion ein. Die Histaminwirkung wird nicht beeinflußt.

Man kann die genannten Stoffe aus der Gefäßwand extrahieren. Bei der Reini-
gung des erschlaffungsfördernden Stoffes wurde in Zusammenarbeit mit LER-
GIER und STUDER festgestellt, daß die Zunahme seiner Aktivität mit der
fortschreitenden Reinheit des Präparates, welches ein Polypeptid darstellt,
parallel geht. Das aktivste Material ergibt nach Totalhydrolyse 15 Amino-
säuren von 10 verschiedenen Typen. Als N-terminale Aminosäure konnte
Arginin nachgewiesen werden. Weitere Arbeiten zur chemischen Identifizie-
rung der beiden Stoffe sind im Gange.

The Effect of Beta-Receptors on the Vascular Escape Phenomenon

By H. Henrich, J. Biester, and J. Lutz
Physiologisches Institut der Universität Würzburg, Germany

The vascular escape-phenomenon (VEP), i.e. the secondary decrease in resistance by a vasoconstriction, is not yet apparent in its formation even in humoral eliciting. In the middle of the well known interpretations are the adrenergic receptors, partly meaning an inequal distribution of the alpha-receptors and a resulting functional shunt of blood (FOLKOW et al., 1964; DRESEL and WALLENTIN, 1966; GREENWAY et al., 1967; a.o.), partly as a secondary persistent effect of the beta-receptors (ROSS, 1967; GEROVA and GERO, 1968).

In the vascular bed of the A. mesenterica sup. in rats, perfused by a constant flow of the rat's blood, there are infused the oligopeptides Vasopressin and Angiotensin and in comparison to them the catecholamines Norepinephrine and Epinephrine. For reactions like the tachyphylaxia, which are similar to the escape-process, care was taken by using low local concentrations and by setting the reactions in couples. Also we considered a small time-dependent decrease in the primary contraction and a conditioned reduction of the escape reaction. The present results show that the VEP also appears in non-catecholamines and consequently it is not dependent on a different localization or efficiency.

The other adrenergic-humoral factor that takes place in the VEP is seen in a beta-receptors effect, which persists longer than the effect of the alpha-receptors. Because of these findings it was necessary to investigate three beta-receptor-blocking substances with different sideeffects regarding their effect on the escape, which is induced by Epinephrine or Norepinephrine (optimal local concentration 1.35 μg/ml). The used beta-blocking drugs are different in their cardioplegity and negative local myotropic behaviour, these properties are primary responsible for the decrease of the initial contraction. The size of the escape process after rising doses of beta-blockers (Propranolol 0.125-1.0; Kö 1366 C.H. BOEHRINGER Sohn, Ingelheim; 0.03-0.25 mg; LB 46; (Viskens) SANDOZ AG.; 0.125-1.0 mg) is enlarged partly in Norepinephrine-reactions, mostly in Epinephrine-reactions depending on the altitude of the initial contraction.

The original recording (fig.) underlines the stated results: After an intraarterial injection of 0.125 mg LB 46 the primary contraction evoced by Norepinephrine is greater than before the beta-receptor-blocking induced reaction. The reagibility of the beta-blockers is controlled by Isoproterenol. After the application of the beta-blocking substances the norepinephrine- and epinephrine- induced escape-reactions are significantly stabilized and quite apparently. These results suggest there is no causal connection between the VEP and the beta-adrenergic effect upon the vascular smooth muscle, because even extreme doses of beta-receptor-blocking substances are unable to suppress the escape. After the VEP appears during use of non-catecholamines, an exclusively adrenergic reason for this appearance would seem improbable.

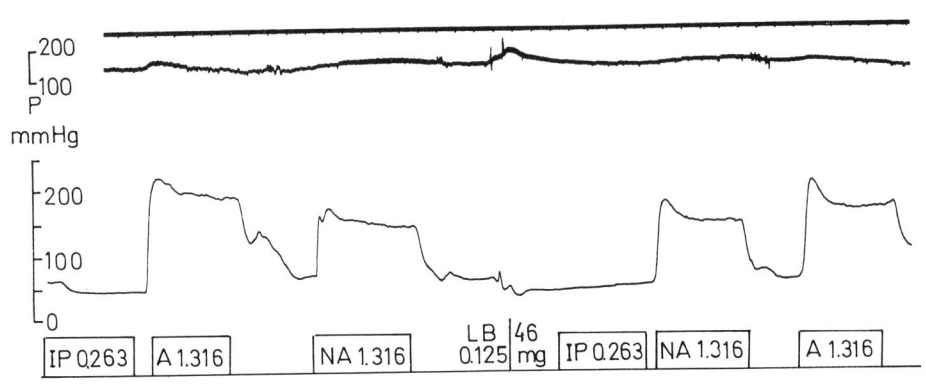

A=Adrenalin
NA=Noradrenalin
IP=Isoproterenol

Lokale Konzentration (µg/ml)
Wistarratte R 163 ♂
Körpergewicht 370 g

References

1) DRESEL, P., WALLENTIN, I.: Acta physiol. scand. 66, 427-436 (1966).
2) FOLKOW, B. et al.: Acta physiol. scand. 61, 445-457 (1964).
3) - Acta physiol. scand. 61, 458-466 (1964).
4) GEROVA, M., GERO, J.: Experientia (Basel) 24, 1134-1135 (1968).
5) GREENWAY, C.V. et al.: J. Physiol. 192, 21-41 (1967).
6) ROSS, G.: Amer. J. Physiol. 212, 1037-1042 (1967).

Possible Mechanisms of Vascular Reactivity Differences in Spontaneously Hypertensive and Normotensive Rat Aortae

By S. Shibata, K. Kurahashi, and J. Mori
Department of Pharmacology, School of Medicine, University of Hawaii,
Honolulu, Hawaii, USA

The major question regarding the functional properties of hypertensive blood vessels is whether they are hyper-reactive or not to excitatory stimuli. Despite numerous studies on the hypertensive animal, the vascular reaction to norepinephrine has received little attention. The results regarding the vascular reactivity of the spontaneously hypertensive rat to norepinephrine has been controversial. The present report describes the results of an attempt to define the differences in the reactivity to certain excitatory stimuli between normotensive and hypertensive rat vascular smooth musculature.

In this experiment, normotensive Wistar rats (NWR: Mean blood pressure, 135 ± 7 mmHg, 300 g) and spontaneously hypertensive rats (SHR: Mean blood pressure 225 ± 7 mmHg, 200-300 g), which were bred by OKAMOTO and AOKI (1), were comparatively studied. The contractile responses of SHR aortae to norepinephrine, potassium, barium and angiotensin were less than that of NWR aortae. Similar results were also obtained when SHR were compared with Strague-Dawley's normotensive rats (140 ± 7 mmHg, 300 g). The total contraction of NWR in response to norepinephrine was differentiated into two components. A fast constituent followed by a slow factor. In contrast, response of the SHR to norepinephrine resulted in showing only the first, fast component, failing to have the second, slow factor of which totalled about 40 % of the NWR contraction. The contraction amplitude of the fast components from both the NWR and the SHR aortae showed no significant difference.

In low Ca^{++} (1/8) medium, NWR failed to elicit the slow component of norepinephrine response. Therefore, in this medium, the dose-response curve of norepinephrine collating aortae responses from both rat types, was not significantly different. Similar contractile curves also resulted with potassium application.

It is interesting to further note that Mn^{++}, Co^{++}, Sr^{++} or La^{+++} caused contraction only in the SHR aorta. There was no responsive difference to excess Ca^{++} in a Ca^{++}-free medium containing potassium. Both aortae failed to show a detectable response to nicotine and tyramine (both 10^{-4} M).

When potassium (10-15 mM), barium (1 mM), epinephrine (10^{-9} M) and serotonine (10^{-7} - 10^{-6} M) were administered to the NWR aortae, a rhythmic contractility was observed. This rhythmicity was not seen in the SHR aortae. This result may be attributed to the multi-unit nature of SHR vascular smooth muscle which is distinguished from the single unit nature of NWR vascular smooth muscle.

These results indicate not only low vascular reactivity of SHR aortic strips to certain excitatory stimuli but also imply that the nature of SHR vascular smooth muscle is physiologically different from that of NWR. As a possible explanation for this difference in reactivity of SHR and NWR aortae, the involvement of a difference in available calcium and its effect on the contractile process is hypothesized.

References

1) OKAMOTO, K., AOKI, K.: Development of a strain of spontaneously hypertensive rats. Jap. Circulat. J. 27, 282 (1963).

Effect of Sympathetic Nerve Stimulation on Cerebral and Cephalic Flow

By C. E. Rapela
Bowman Gray School of Medicine, Winston-Salem, N. C. USA
(read by H. D. Green)

Previous work (RAPELA et al., Circulat. Res. 21, 559-568, 1967) indicated that the carotid sinus reflex has no influence on the cerebral vascular tone and on the autoregulatory responses of cerebral blood flow to changes in perfusion pressure. Also, most workers have observed that the cervical sympathetic has none or slight effect on the cerebral vasculature. Recently (D'ALECY and FEIGL, Fed. Proc. 29, 520, 1970) it has been postulated that stimulation of the stellate ganglion, or nerves arising from it to the superior cervical sympathetic ganglion of the dog, produces marked constriction of the cerebral vasculature.

We have studied the effect of stimulating the stellate ganglion and its nerve branches on the canine cerebral and cephalic blood flow. The cerebral venous blood flow was measured at the confluence of sagittal, straight and lateral sinuses with the lateral sinuses occluded (RAPELA et al., Fed. Proc. 20, 100, 1961). The blood flow in the common carotid ipsilateral to the nerve stimulated was measured with a non-cannulating electromagnetic flowmeter. All branches from the stellate ganglia were isolated and prepared for stimulation with a bipolar electrode. Monophasic pulses of 10 msec duration and 3 volts were applied at a frequency of 14 per sec during 1 min. Stimulation of the stellate ganglion or the nerve branch to the superior cervical sympathetic ganglion induced a marked decrease of common carotid blood flow and dilation of the ipsilateral pupil but no change in the cerebral vascular conductance.

It is concluded that electrical stimulation of the cervical sympathetic has no effect on the cerebral vasculature. Effects of sympathetic stimulation on measured cerebral blood flow are observed only if significant communications between intra- and extracranial venous vasculature are present.

Supported by NIH grant 487 and North Carolina Heart Association.

Potentiation of Epinephrine Response in Vascular Tissue by Nicotine

By A. Maiti
Department of Biochemistry, University College of Medicine, Calcutta, India

In studies of the influence of nicotine on the peripheral adrenergic neuroeffector transmission in the vascular smooth muscles, experiments were designed
to analyse the vascular reactivity towards epinephrine following nicotine treatment. Topical or systemic administration of nicotine caused constriction of
arterioles, metarterioles and precapillaries along with suppressed vasomotion in the peripheral microvasculature of rat mesocecum. Comparably, it
also induced a transient strong contraction of isolated spirally cut cat aortic
strip in vitro preparations. Such responses of nicotine were gradually decreased upon its repeated administration, but exhaustive washing and resting
restored the reduced responses. While the tachyphylactic response of nicotine was pronounced, exposure of the above preparations to test doses of epinephrine not only restored the initial responses but the magnitude of such
responses were greatly potentiated. On the other hand, nicotine produced
tachyphylaxis depressed the responses to serotonin, histamine, amphetamine
and tyramine. It appeared that repeated nicotine pretreatment modulated the
pattern of microcirculatory function and vascular smooth muscle responsiveness to epinephrine possibly through an alteration in the affinity constant for
catecholamines at the terminal adrenoceptor of the effector cells.

Comparison between Sympathetic Activity in Human Skin- and Muscle Nerves and Accompanying Changes in Vascular Resistance

By W. Delius, B. G. Wallin, A. Hongell, and K.-E. Hagbarth
Laboratory of Clinical Neurophysiology and Departments of Internal Medicine
and Clinical Physiology, University Hospital, Uppsala, Sweden

The results of numerous investigations of the peripheral circulation indicate
a highly differentiated control of the sympathetic outflow to different vascular beds. Neurophysiological evidence for differentiated control of sympathetic outflow to arterial muscle- and skin vessels has now been given (HAG-
BARTH and VALLBO, 1968; WALLIN, DELIUS, HONGELL and HAGBARTH,
1971 a, b, d; WALLIN, HONGELL, HALLIN, TOREBJÖRK and HAGBARTH,
1971 c). Multi-unit sympathetic activity was recorded by microelectrodes

inserted percutaneously into muscle- and skin nerves in alert, adult subjects together with measurements of intraarterial blood pressure and muscle- and skin blood flow in the extremities. The nervous signals were abolished by sympathetic ganglion blocking agents and by Lidocaine nerve blocks proximal to the recording site.

Spontaneous sympathetic activity has different characteristics in human muscle- and skin nerves. The activity recorded from muscle nerves consists of bursts of impulses occuring pulse synchronously in short irregular sequences, separated by periods of relative silence. The impulses cause vasoconstriction and their occurence is intimately related to spontaneous blood pressure variations, suggesting an inhibitory baroreflex influence on the sympathetic outflow. The sympathetic activity led off from skin nerves, on the other hand, appear in irregular bursts of varying duration showing neither pulse synchronous grouping nor correlation to blood pressure fluctuations.

When comparing neural and effector organ responses during different manoeuvres the changes in sympathetic activity correlated well with changes in skin vascular resistance. Manoeuvres which are known to elicit baroreflex responses caused changes in sympathetic outflow to skeletal muscles but not to the skin. The reverse was seen in response to thermal stimuli. During mental arithmetic we found opposite sympathetic responses in skin and muscle nerves, corresponding to the well-known simultaneous constriction in skin and dilatation in muscle vessels. Muscle work was regularly accompanied by an increased sympathetic outflow to the muscles not engaged in the work and an increase in muscular vascular resistance, whereas the sympathetic activity to the skin was either uninfluenced or slightly increased.

The high vascular resistance in skeletal muscle in spite of complete cessation of the sympathetic outflow during the relaxation phase after the Valsalva manoeuvre is discussed as an example of apparent disagreement between neural and vascular events.

References

1) HAGBARTH, K.-E., VALLBO, A.B.: Pulse and respiratory grouping of sympathetic impulses in human muscle nerves. Acta physiol. scand. 74, 96-108 (1968).
2) WALLIN, B.G., DELIUS, W., HONGELL, A., HAGBARTH, K.-E.: General characteristics of sympathetic activity in human muscle nerves. (to be published) 1971 a.
3) ---- Manoeuvres affecting sympathetic outflow in human muscle nerves. (to be published) 1971 b.
4) WALLIN, B.G., HONGELL, A., HALLIN, R., TOREBJÖRK, E., HAGBARTH, K.-E.: General characteristics of sympathetic activity in human skin nerves. (to be published) 1971 c.
5) WALLIN, B.G., DELIUS, W., HONGELL, A., HAGBARTH, K.-E.: Manoeuvres affecting sympathetic outflow in human skin nerves. (to be published) 1971 d.

Further Evidence for the Existence of Different Types of Angiotensin II Receptors

By A. Papadimitriou and M. Worcel

Centre de Recherches sur l'Hypertension Artérielle, Hôpital Broussais, Paris
France

Formerly (MEYER, PAPADIMITRIOU and WORCEL, 1970),we have shown some results of a study of the action of five different structure analogues of Valyl-5-angiotensinamide II on the contractility of three smooth muscle preparations. The biological responses of angiotensin and the analogues were then compared by finding the concentration of each that was required in the tissue bath to cause an equal and moderate response of the muscle (at around the ED 50 of Valyl-5-angiotensinamide).

The problem was further studied by analyzing the activity of three other analogues. As previously the organs tested were isolated, rat colon and uterus and strips of rabbit aorta suspended in a temperature regulated, oxygenated Krebs solution. In each experiment, complete dose-response curves were obtained for Valyl-5-angiotensinamide and one of the analogues alternatively. This procedure served to determine the ED 50 and the intrinsic activity (ARIENS, SIMONIS and van ROSSUM, 1964) of each of the analogues used.

The ratios of ED 50 and intrinsic activities are variable from organ to organ. Theoretically, if the receptors in the three different organs were the same, similar ratios of activities of the analogues compared with Valyl-5-angiotensinamide II would be expected. The results indicate that the angiotensin receptors are different in the three organs tested.

References

1) ARIENS, E. J., SIMONIS, A. M., van ROSSUM, J. M.: Drug-receptor interaction: Interaction of one or more drugs with one receptor system. In: Molecular Pharmacology, E. J. Ariens (Ed.). New York - London: Academic Press, 1964, pp. 119-286.
2) MEYER, P., PAPADIMITRIOU, A., WORCEL, M.: Possible existence of different types of angiotensin II receptors. Brit. J. Pharmacol. 40, 543P (1970).

All the analogues were obtained by the courtesy of Dr. B. Riniker (CIBA Research Laboratories, Basle), and S. Fernandjean (C. E. A., Saclay).

Individualization of Angiotensin Receptors in Smooth Muscle Cell

By M. Baudouin and P. Meyer
Hypertension Laboratory, Broussais Hôpital, Paris, France

It is generally accepted that the first step of vasoconstriction induced by vasoactive hormones is represented by interaction of the vasoactive agent with
some specific cellular component. The direct demonstration of this specific
receptor requires the availability of a radioactive hormone possessing the
full biological activity of the unlabelled hormone, and the kinetics of binding
of the radioactive tracer should be similar to those of the biological response
Using a highly tritiated angiotensin of specific activity of 56 Ci per mMole
(1) we were able to demonstrate in rabbit aorta a specific binder for angiotensin II fullfilling these criteria, and to locate it in microsomal fraction. The
first part of this study was performed on whole rabbit aortae dissected free of
adventitia (2). Bound radioactivity after exposition to ^{3}H-angiotensin II was
determined in hydrolyzed tissue by substracting from total tritium content
contained in extracellular fluid volume measured by ^{14}C Insulin.

The fundamental characteristics of a specific binding system were demonstrated. 1) Reversibility was shown by the time courses of binding of radioactivity, and by washing the aorta with hormone-free medium after incubation. The kinetic constants of uptake and release were reduced by cooling the
medium at + 4 $^{\circ}$C.
2) High affinity of the aorta for angiotensin was demonstrated by the lack of
binding of radioactivity in non-target tissues.
3) Limited capacity was inferred from the results of radioactivity binding at
different concentrations of ^{3}H-angiotensin II in the medium. A S-shaped curve exhibiting the saturation kinetics of a specific system was observed from
10^{-9} to 10^{-7} M. For concentrations of ^{3}H-angiotensin above 10^{-7} M, the proportion of bound radioactivity varied directly with the concentration of ligand,
suggesting non-specific binding. This assumption has been verified by the
proportion graph method using a 360 IBM computer. The binding parameters
of the specific system obtained by computerized analysis were K = 13.1 $\pm$
7.7 nM^{-1} and N = 8.4 $\pm$ 4.8 $\cdot$ 10^{-15} M $\cdot$ mg^{-1}.
4) Specificity of binding was demonstrated by a) time courses of binding of
radioactivity and of the contractile response were identical, b) the kinetic
constants of the specific binding and the biological response were also similar, c) bound radioactivity was found to be essentially represented by intact
angiotensin, d) inhibition of bound radioactivity by unlabelled angiotensin II,
3-8 hexapeptide and angiotensin analogues substituted in positions 2, 4, 6 and
8, was proportional to the respective biological activities of these derivates.
These observations strongly indicated that the saturable binding sites demonstrated here may be those involved in the biological response.

The next step of the study (3) was performed on microsomal vesicles prepared from rabbit aorta by differential centrifugations in sucrose between 5000

and 102 000 g. These vesicles may derive from all membrane, endoplastic reticulum, and essentially micropinocytotic vesicles. No cytochrome C oxydase activity was detected demonstrating the absence of mitochondria contamination. In presence of ATP, these vesicles were able to bind calcium. This was demonstrated with the use of 45 Ca, the bound calcium on vesicles being measured by the millipore filtration technique. In presence of 100 mM KCl, 5 mM MgCl2 and 5 mM ATP, calcium binding reached equilibrium after 10 minutes. Calcium binding was saturated at a calcium concentration of 10^{-4} M in the medium, the calcium binding capacity being 6.5 pMoles per /ug of protein.

The release of incorporated calcium was studied by diluting the incubation medium in 25 volumes of calcium-free medium, and determining the calcium remaining in the vesicles by the filtration technique at different time intervals. Half of the incorporated calcium was released into the medium within 1 minute. Angiotensin increasingly affected calcium release both in presence and in absence of ATP to a significant effect.

The releasing effect induced by angiotensin is dose-dependent, and reaches a maximum at concentrations of angiotensin II in the medium corresponding to a 25 % increase in the rate of release occuring in the absence of the hormone.

The binding of ^{3}H-angiotensin to the vesicles was simultaneously studied after an incubation of 2 minutes, the separation of bound and unbound radioactivity being performed by means of dextran coated charcoal. Saturation of binding sites also occured in the range of 1 to 3.2 x 10^{-7}M.

The binding parameters found in the vesicles are very close to those of the specific system found on whole aorta. A good relationship was found between the binding parameters and the parameters of biological responses both in aorta and in vesicles.

The present investigation therefore demonstrates that specific angiotensin receptors are located in the microsomal fraction and that a relative permeability change in calcium results from hormone-reception interaction. The resulting increase in cytosolic calcium concentration may account for the contraction of the myofilament.

References

1) MORGAT, J.L., HUNG, L.T., FROMAGEOT, P.: Biochim. biophys. Acta 207, 374 (1970).
2) BAUDOUIN, M., MEYER, P., WORCEL, M.: Biochem. biophys. Res. Commun. 42, 434 (1971).
3) BAUDOUIN, M., MEYER, P.: In preparation.

Diskussion

JOHANSSON: Es gibt Fälle von Übersensibilität auf exogenes Noradrenalin. Er glaubt jedoch nicht, daß so etwas im Falle der Befunde von TÖREK zutreffe. TÖREK antwortet, daß die Übersensibilität auf exogenes Noradrenalin ein praesynaptischer Effekt sei. GODFRAIND fragt nach der Zunahme des Norepinephringehalts in der Muskelzelle in Beziehung zur Zeit. TÖREK antwortet, daß die tunica media nach 1 Minute 20 % des Gehalts der Umgebungsflüssigkeit enthält, nach 2 Minuten 50 %.

HORN fragt, ob nur Noradrenalin oder ob auch Adrenalin verwandt wurde, da zwischen diesen beiden Substanzen ein Unterschied in der Aufnahme durch die Nervenendigungen besteht.

BEVAN antwortet, daß er nur Noradrenalin verwandt habe, weil nach seiner Auffassung nur Noradrenalin eine Überträgersubstanz sei.

HORN weist darauf hin, daß das Escape-Phänomen nicht nur bei Katecholamin stattfindet, bei denen es FOLKOW und Mitarb. gefunden haben, sondern auch bei anderen vasoaktiven Substanzen. Er fragt, wie lange die anderen vasoaktiven Substanzen gegeben wurden.

HENRICH antwortet: die Reaktionszeit betrug 5 Minuten und die Primärkontraktion trete in der Regel nach 30 Sekunden ein.

Zum Vortrag RAPELA weist LASSEN darauf hin, daß bei der Messung der Gehirndurchblutung pro Gramm/Gewebe das Gewicht des Gehirns bekannt sein muß. Bei Messungen des Ausstroms aus dem Venensinus besteht die Möglichkeit, daß der Flow nicht direkt sondern die Differenz zwischen Atmosphärendruck und dem Druck im Venensinus erfaßt wird.

In der Erwiderung weist GREEN darauf hin, daß RAPELA die Durchblutung durch Einstrom-Messungen in einem Bereich der Carotis registriert hat und daß außerdem bei Verschluß der Vene der Druck im Sinus venosus bis auf arterielle Werte erhöht wurde, was eindeutig dafür spricht, daß in der Tat eine Durchblutungsmessung vorgenommen worden ist. Außerdem lagen die Werte der corticalen Durchblutung etwa in den Meßbereichen, in denen auch sonstige mit anderen Verfahren gewonnene Meßwerte zu finden waren. HIRSCH weist darauf hin, daß HYDEN bei einer Reihe von Reizungen des sympathische Systems Erhöhungen der Durchblutung gefunden hat. Allerdings waren das nur Einzelfälle.

In der Antwort sagt GREEN, daß bei Bestehen der Autoregulation, und das sind die Mehrzahl der Fälle, derartige Erhöhungen nicht vorkommen können, während bei Aufheben der Autoregulation etwa unter hoher CO_2-Konzentration in der Tat blutdruckpassive Steigerungen der Gehirndurchblutung möglich sind. Diese allerdings bei ausgeschaltetem Gefäßtonus.

KOEPCHEN fragt DELIUS, ob die Blutdruckänderungen eine Veränderung der sympathischen Aktivität bedingen oder ob primär eine Änderung der sympathischen Aktivität zur Blutdruckveränderung führt.

Bei der Antwort auf diese Frage werden Untersuchungen aus der THAUER' schen Arbeitsgruppe (Bad Nauheim) erwähnt, die Hinweise darauf geben, daß primär eine sympathische Aktivität an der Änderung der glatt-muskulären Gefäßstrukturen den Druck verändern könne.

Zum Vortrag PAPADIMITROU und WORCEL fragt PEIPER, was wohl die Ursache dafür sei, daß in bovinen Gefäßen Angiotensin keinerlei Wirkung zeige, während bei der Ratte eine sehr starke Kontraktion auftrete. Antwort: Die Ursache dafür sei in den Rezeptoren zu suchen. Manche Gefäße hätten sie, andere nicht. Bei der Diskussion der Frage, ob irgendwelche quantitativen Angaben darüber zur Calciumfreisetzung bei der Aktivierung der kontraktiven Proteine vorliegen, wird geantwortet: Wir haben eine ungefähre Vorstellung, daß etwa das Tausendfache der intrazellulären Konzentration zu einer Kontraktion führt.

Frage von GODFRAIND: Haben Sie irgendwelche Hinweise für die Wirkung von verschiedenen ionalen Zusammensetzungen auf die Freisetzung des Calciums? Antwort: Anstiege des Natriums verursachten eine signifikante Verminderung der Inkorporation von Calcium in den Vesikeln.

Chairmen: U. Peiper, A. Hirsch

Local Vascular Effects of Potassium and Magnesium Depletion and the Role of Potassium in Active Hyperemia

By D. K. Anderson, R. A. Brace, S. A. Roth, D. P. Radawski, J. B. Scott, and F. J. Haddy
Departments of Physiology and Chemical Engineering, Michigan State University, East Lansing, Michigan, USA

Hemodialysis was used to study the effect of hypokalemia and hypomagnesemia on the resistance to blood flow through the gracilis muscle of the dog and to examine the role of potassium in exercise hyperemia. When the $[K^+]$ of the perfusing blood was decreased (or increased) rapidly, there was a corresponding increase (or decrease) in the vascular resistance. Hypomagnesemia had no apparent effect on the resistance, either alone or in combination with low potassium. The results are summarized on fig. 1, which shows the percent change in perfusion pressure at constant flow when $[K^+]$ and/or $[Mg^{++}]$ levels of the arterial blood are raised or lowered. Each datum is the average of several changes in a single animal.

Depletion of 10 percent of the muscle potassium by prolonged (1-2 hrs) dialysis had no effect on active hyperemia elicited by nerve stimulation. Stimulation raised venous potassium levels by approximately 2 mEq/liter (from 3.9 to 5.6 mEq/liter when perfusing with normal blood and from 1.5 to 3.4 mEq/liter when perfusing with hypokalemic blood). However, the resistance change was the same in each case. Thus there was no correlation between the absolute level of venous potassium and the resistance change during exercise. Lowering inflow $[K^+]$ during exercise dilation did not change the resistance. Finally, increasing venous $[K^+]$ in the resting gracilis muscle (by dialyzing the arterial blood) to levels seen during exercise failed to produce comparable decreases in vascular resistance. These findings indicate (1) hypokalemia increases resistance to blood flow, (2) hypomagnesemia has no effect and (3) while potassium may be a factor in the dilation mechanism during active hyperemia, other factors must be involved.

The authors gratefully acknowledge the financial support of the Michigan Heart Association.

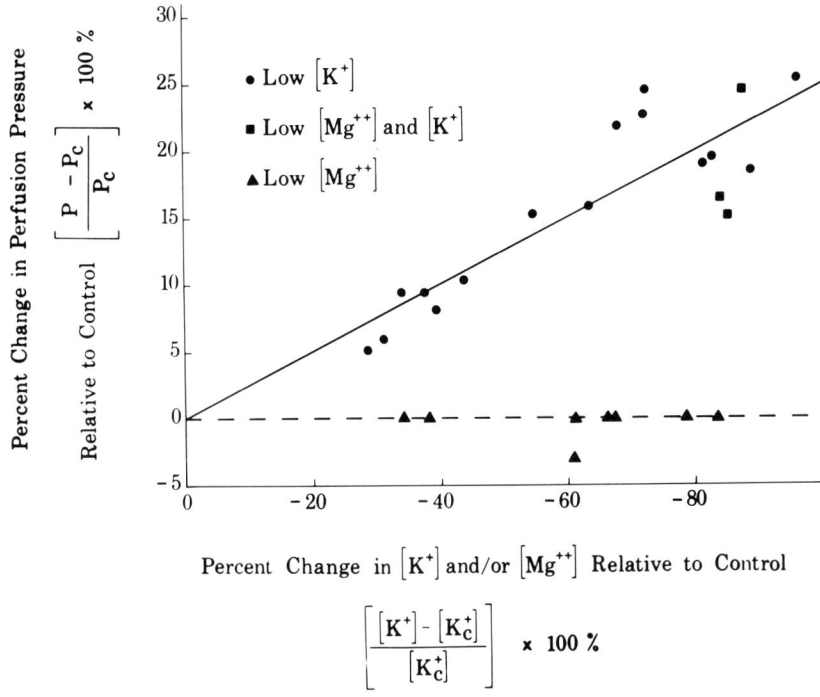

Evidence that Na Rather Than Ca Ions Carry the Depolarizing Current of Arterial Action Potentials

By J. M. Graham and W. R. Keatinge
The London Hospital Medical College, University of London, Department of
Physiology, London, Great Britain

Intestinal smooth muscle usually ceases electrical activity in Ca-free solutions, while continuing it in Na-free solutions containing Ca. However, sheep carotid arteries develop electrical activity in Ca-free solutions (KEATINGE, 1968 a). This activity was stopped when Na was replaced by Tris or choline and was associated with an increased influx of Na (KEATINGE, 1968 b). Tetrodotoxin did not inhibit electrical activity of the arterial smooth muscle either in normal or Ca-free solutions.

The chelating agent EDTA may be added to exclude the possibility that trace amounts of Ca present in Ca-free solutions are sufficient to sustain Ca-based activity, but in Ca-free solutions at 36 ^{O}C this stops spike activity within a few minutes by causing extreme depolarization of the preparation.

Sucrose-gap recordings have now been obtained from strips of sheep carotid artery showing that at 0-10 ^{O}C electrical activity persists for long periods in the presence of Ca-free solutions containing a high concentration of EDTA. In Ca-free solution containing EDTA 12.5 mM at 0-10 ^{O}C electrical activity developed and at such low temperatures the muscle did not become highly depolarized and spikes continued for as long as two hours. This activity was also found to be insensitive to Tetrodotoxin 10^{-5} M; in some cases Tetrodotoxin even caused a small increase in the amplitude of the spikes. In most other tissues with Na-based electrical activity Tetrodotoxin blocks Na influx, but there is at least one other exception in the puffer-fish nerve (KAO, 1966).

References

1) KAO, C.Y.: Pharmacol. Rev. 18, 997-1049 (1966).
2) KEATINGE, W.R.: J. Physiol. 194, 169-182 (1968 a).
3) - J. Physiol. 194, 183-200 (1968 b).

Prinzipielles zur Wirkung von Ca^{++}-Antagonisten auf die bioelektrische und mechanische Funktion glatter Muskelzellen

Von A. Fleckenstein und G. Grün
Physiologisches Institut der Universität Freiburg i. Br., Germany

Der Wirkungsmechanismus von sogenannten "muskulotropen" Relaxantien der glatten Muskulatur war bislang nicht genügend physiologisch definiert. Eigene Versuche (1,2) haben jedoch kürzlich gezeigt, daß einige typische Vertreter aus dieser Gruppe die elektromechanischen Koppelungsprozesse durch einen Ca^{++}-antagonistischen Effekt blockieren können. Infolgedessen werden die mechanischen Reaktionen der glatten Muskulatur bei Membran-Depolarisation durch KCl, Katelektrotonus u.a. unterdrückt, ohne daß dabei das Ausmaß der Potentialveränderungen an der Membran stärker eingeschränkt wird. Darüberhinaus können solche Stoffe an glatten Muskelzellen auch die elektrische Spontan-Aktivität noch in sehr geringer Konzentration zum Stillstand bringen. Extra-Calcium (7,2 mM/l) restituiert üblicherweise die Fähigkeit zur Kontraktur und zur automatischen Spike-Produktion, während Ca^{++}-Mangel die Ca^{++}-antagonistischen Hemm-Effekte dieser Substanzen potenziert.

Tab. 1. Ca^{++}-antagonistische Inhibitoren der elektromechanischen Koppelung am Warmblüter-Myokard mit gleichzeitig "muskulotropen" Relaxationseffekten an der glatten Muskulatur.

Segontin (Prenylamin) Farbwerke Hoechst	
Isoptin (Verapamil) (Iproveratril) Knoll AG	
Substanz D 600 Knoll AG	
Substanz Bay a 1040 Bayer-Werke	

Als Prototyp "muskulotroper" Relaxantien gilt bekanntlich Papaverin. Etwa 5-100 mal stärker sind allerdings die - ursprünglich am Warmblüter-Myokard studierten - spezifischen Ca^{++}-Antagonisten Verapamil (Isoptin), D 600 (ein Methoxyderivat von Verapamil) sowie Prenylamin (Segontin) (3,4). Der Spitzenstoff in dieser Reihe ist jedoch die neue Substanz Bay a 1040 (vgl. Tab. 1). Zur Ausschaltung der Spontan-Aktivität des isolierten Uterusgewebes von Mensch und Ratte sowie zur Relaxation kontrahierter Gefäßstreifen von Coronararterien (vgl. Abb. 1) und Aorten genügt hier schon eine Konzentration von 0,05 mg/l. Zur elektromechanischen Entkoppelung sind an der Uterus-Muskulatur meist etwa 10-fach höhere Dosen notwendig. Die Blockierung der Bildung und Fortleitung von Aktionspotentialen in glatten Muskelzellen durch die genannten Ca^{++}-Antagonisten scheint in Übereinstimmung mit BÜLBRING und TOMITA (5) auf die Möglichkeit einer stärkeren Beteiligung der Ca^{++}-Ionen an den Erregungsprozessen der glatten Muskelzellen hinzuweisen.

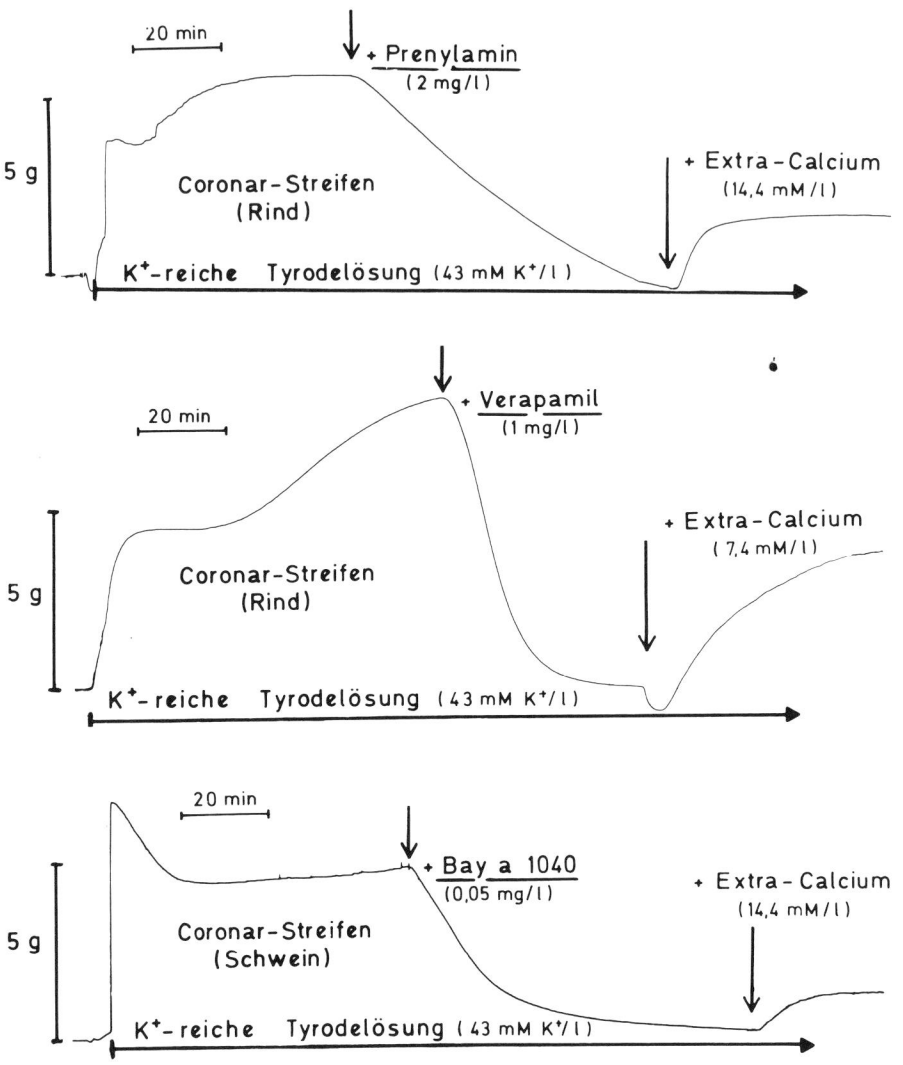

Abb. 1. Aufhebung der Kalium-Kontraktur isolierter Coronarstreifen durch Ca^{++}-antagonistische Coronardilatatoren. Partielle Restitution der Kontraktilität durch Extracalcium. Kontraktur-Erzeugung mittels isotonischer K^{+}-reicher Tyrodelösung, enthaltend 43 mM K^{+}/l (unter Reduktion von Na^{+}) und 1 mM Ca^{++}/l. Temperatur 35 $^{\circ}$C. Die Aktivierung des kontraktilen Systems in den glatten Gefäßmuskelzellen erfolgt bei dieser Versuchsanordnung wohl in erster Linie durch den Einstrom von extracellulären Ca^{++}-Ionen durch die K^{+}-depolarisierte Membran ins Zellinnere. Diese transmembranären Ca^{++}-Bewegungen werden durch die genannten Ca^{++}-Antagonisten offenbar ähnlich wie an Myokardfasern unterdrückt

References

1) GRÜN, G., FLECKENSTEIN, A., TRITTHART, H.: Naunyn-Schmiedeberg's Arch. Pharmak. 264, 239 (1969).
2) FLECKENSTEIN, A., GRÜN, G., TRITTHART, H., BYON, K.: Klin.
   Wschr. 49, 32-41 (1971).
3) FLECKENSTEIN, A., DÖRING, H.J., KAMMERMEIER, H.: Klin. Wschr.
   46, 343-351 (1968).
4) FLECKENSTEIN, A., TRITTHART, H., FLECKENSTEIN, B., HERBST,
   A., GRÜN, G.: Pflügers Arch. ges. Physiol. 307, R25 (1969).
5) BÜLBRING, E., TOMITA, T.: J. Physiol. 196, 137P (1968).
6) BÜLBRING, E., BRADING, A.F., JONES, A.W., TOMITA, T.: Smooth
   muscle. London: Edward Arnold Ltd. 1970.

Contraction and Calcium Pools Mobilisation in Rat Aorta

By T. Godfraind and A. Kaba
Laboratoire de Pharmacodynamie Generale, Universite de Louvain, Belgium

Calcium appears to act as an intracellular transmitter for excitation-contraction coupling. As far as vascular smooth muscle is concerned, there is an obvious relation between the extracellular calcium concentration and the force of the tonic tension developed in the presence of adrenaline or in an isotonic potassium chloride solution. It has been postulated (DANIEL, 1965) that smooth muscle calcium could be distributed among a labile and a sequestered pool and there are many conflicting opinions regarding the possible role of each of these pools in drug action (GODFRAIND, 1971; GODFRAIND and KABA, 1969; SOMLYO and SOMLYO, 1968). The role of these calcium pools was investigated by the analysis of rat aorta contraction in the absence and in the presence of calcium antagonists and by the study of calcium exchange in resting and stimulated preparations.

Experiments here summarized were performed in steady state conditions at 37 °C on helical strips of rat aorta stretched isometrically at 1 g. Bathing solutions were either a physiological solution (m-moles: NaCl 122, $NaHCO_3$ 15, KCl 5.9, $CaCl_2$ 1.25, $MgCl_2$ 1.25 and glucose 11) or a high potassium solution (m-moles: NaCl 27, $NaHCO_3$ 15, KCl 101, $CaCl_2$ 1.25, $MgCl_2$ 1.25 and glucose 11).

In response to adrenaline or in the presence of high potassium solution, rat aorta shows a phasic and a tonic contraction (fig. 1). There are several facts which show that the phasic contraction induced by adrenaline is not controled by the same factors as that induced by high potassium solution. As fig. 1 illustrates, the dissociation of the two parts of the contractile response was more marked with adrenaline. Furthermore, the duration of the incubation in a calcium free solution necessary to abolish the phasic contraction was shorter for high potassium than for adrenaline. Finally, calcium antagonists such as cinnarizine or chlorpromazine affected differently this contraction (GODFRAIND, 1971; GODFRAIND and KABA, 1969 a). On the basis of these observations confirming experiments performed with rabbit mesenteric arteries (GODFRAIND and KABA, 1969 b), we postulated that adrenaline phasic contraction was due to the release of sequestered calcium. This hypothesis was tested by the study of calcium exchange (GODFRAIND and KABA, 1971).

The calcium content of rat aorta was equal to 3.76 ± 0.09 (n = 16) m-moles/k wet wt after two hours in physiological solution. The treatment by potassium did not change this content, but adrenaline 10^{-4} M induced a small decrease. In ^{45}Ca physiological solution, there was no further uptake of ^{45}Ca after the

---

This work has been supported by a grant from the F. R. S. M.

first 20 minutes and about 64 per cent of tissue calcium was exchanged for ^{45}Ca.

The ^{45}Ca distribution was studied by the analysis of ^{45}Ca efflux curves. In physiological solution, ^{45}Ca loss presented a complex time course which was divided into slow, fast and very fast components. The slow component had a half-time of 26 minutes and an intercept at zero time of 0.088 m-moles ^{45}Ca/ kg wet wt. The fast loss had a half-time of 2 minutes and an intercept of 1.056 m-moles ^{45}Ca/kg. The very fast occured in the first seconds and corresponded to 1.400 m-moles ^{45}Ca/kg. The ^{45}Ca in solution in the ^{14}C-inulin-space was equivalent to 0.483 m-moles ^{45}Ca/kg. On the basis of these findings, three pools for ^{45}Ca distribution were considered: They were characterized by their rate of exchanging calcium, these pools representing the labile calcium pool.

In high potassium solution, there was a redistribution of ^{45}Ca among these pools, ^{45}Ca being mainly accumulated in the fast exchanging pool which increase in capacity was due to the reduction of the content of the very fast exchanging pool. This change was attributed to the accumulation of ^{45}Ca on Na-Ca sites.

In the presence of adrenaline, the redistribution of ^{45}Ca was qualitatively similar to that observed in high potassium solution, but there were quantitative differences.

The present results suggest that stimulating agents can be characterized by their ability to produce a labilization and a redistribution of cellular calcium.

References

1) DANIEL, E.E.: In muscle.W.M. Paul, E.E. Daniel, L.M. Kay and G. Monckton (Eds.) Oxford: Pergamon Press 1965, pp 295.
2) GODFRAIND, T.: Prod. pharm. 26, 117 (1971).
3) GODFRAIND, T., KABA, A.: Arch. int. Pharmacodyn. 178, 488 (1969 a).
4) GODFRAIND, T., KABA, A.: Brit. J. Pharmacol. 36, 549 (1969 b).
5) GODFRAIND, T., KABA, A.: J. Physiol. 1971 (in press).
6) SOMLYO, A.P., SOMLYO, A.V.: Pharmacol. Rev. 20, 197 (1968).

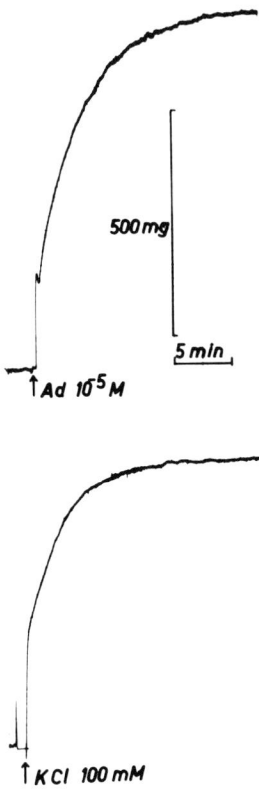

500 mg

5 min

↑ Ad 10⁻⁵ M

↑ KCl 100 mM

Isolated rat aorta, isometric contraction in response to adrenaline (Ad) and
to high potassium solution (KCl)

Blockierung der Ca++-Effekte auf Tonus und Autoregulation der glatten Gefäßmuskulatur durch Ca++-Antagonisten (Verapamil, D 600, Prenylamin, Bay a 1040 u. a.)

Von G. Grün, A. Fleckenstein und K. Byon
Physiologisches Institut der Universität Freiburg i. Br., Germany

Der Tonus der glatten Gefäßmuskulatur wird bei Zunahme der extracellulären Ca^{++}-Konzentration beträchtlich erhöht, während Ca^{++}-Mangel zu Relaxation führt (1, 2). Auch die autoregulatorische Vasokonstriktion bei schrittweiser Steigerung des intravasalen Drucks ist nach unseren Beobachtungen an isoliert perfundierten Kaninchen-Ohren stark Ca^{++}-abhängig: Ca^{++}-Mangel unterdrückt die Autoregulation, Extra-Calcium führt dagegen zu übersteigerten mechanischen Reaktionen (5). Eine pH-Erhöhung wirkt dabei potenzierend auf die Ca^{++}-Effekte, pH-Erniedrigung löscht die Ca^{++}-abhängige Autoregulation aus. Durch α-Rezeptoren-Blockade mittels Phentolamin (Regitin) werden die Ca^{++}-Wirkungen auf Tonus und Autoregulation nicht gehemmt. Ein neuer Weg zur Blockierung der Ca^{++}-Effekte an der glatten Gefäßmuskulatur ist durch die Entdeckung der spezifisch Ca^{++}-antagonistischen Eigenschaften einer Reihe von Pharmaka wie Verapamil, D 600 - ein Methoxyderivat von Verapamil - Prenylamin u. a. eröffnet worden (3, 4). Noch stärker wirkt die erst kürzlich synthetisierte Substanz Bay a 1040 (Abb. 1).

Schon geringe Dosen dieser Verbindungen (1 mg/l und weniger) verursachen eine starke Vasodilatation und beseitigen die Autoregulation. Extra-Calcium wirkt restituierend. Kontrahierte Streifen-Präparate aus Aorten von Kaninchen oder aus Coronarien von Schweinen und Rindern reagieren ebenfalls auf Ca^{++}-Antagonisten mit einem Tonus-Verlust. Die Catecholamin-bedingte Vasokonstriktion wird durch Ca^{++}-Antagonisten weniger stark gehemmt (6). Der Ca^{++}-Antagonismus der genannten Verbindungen verdient als neues vasodilatatorisches Prinzip auch in der Praxis hohe Beachtung, da hierdurch der Fundamental-Prozess der Vasokonstriktion ausgeschaltet werden kann.

Literatur

1) BRECHT, K., ESTADA, J., GÖTZ, A.: Pflügers Arch. ges. Physiol. 279, 330-340 (1964).
2) BOZLER, E.: Amer. J. Physiol. 216, 671-674 (1969).
3) FLECKENSTEIN, A., DÖRING, H.J., KAMMERMEIER, H.: Klin. Wschr. 46, 343-351 (1968).
4) FLECKENSTEIN, A., KAMMERMEIER, H., DÖRING, H.J., FREUND, H.J. unter Mitarb. von G. GRÜN und A. KIENLE: Z. Kreisl.-Forsch. 56, 716-744 (1967).
5) GRÜN, G., FLECKENSTEIN, A.: Naunyn-Schmiedeberg's Arch. Pharmak. Suppl. 270, R48 (1971). 12. Frühjahrstagung d. Dtsch. Pharmakol. Ges. am 22. März 1971 in Mainz.
6) HAEUSLER, G.: Vortrag anlässlich der Tagg. d. Dtsch. Pharmakol. Ges. September 1970 in Heidelberg.

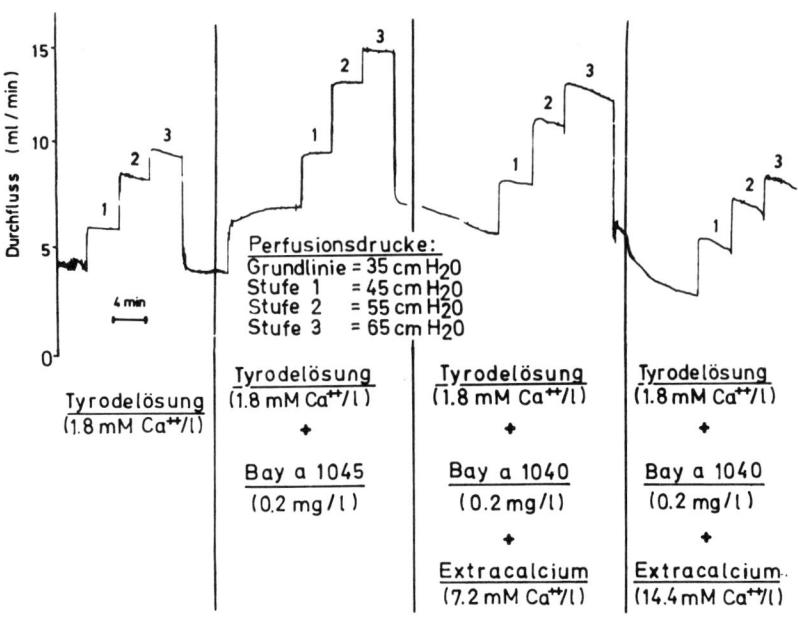

Aufhebung von Gefäßtonus und Autoregulation an einem isoliert durchström-
ten Kaninchenohr durch Bay a 1040 (0, 2 mg/l). Anschließende Restitution de
Gefäßreaktionen durch Extracalcium. In dem Versuch wurde mehrmals hin-
tereinander am gleichen Kaninchenohr eine schrittweise Steigerung des Per-
fusionsdrucks vorgenommen und die hieraus resultierende Zunahme des
Durchflußvolumens mit dem Ratemeter nach KRAUSE (Z. Kreisl.-Forsch.
53, 725 (1964)) gemessen. Im linken Abschnitt des Bildes kommt es bei Perfusion mit gewöhnlicher Tyrodelösung infolge der stufenweisen plötzlichen
Drucksteigerungen um jeweils 10 cm H_2O-Säule zu einer reaktiven autoregulatorischen Vasokonstriktion, sodaß die Ausstrom-Volumina von den gedehn-
ten Ohrgefäßen selbst bei den Druckstufen 2 und 3 wieder partiell gedrosselt
werden. Setzt man der Tyrodelösung den Ca^{++}-Antagonisten Bay a 1040 (0, 2
mg/l) zu, so tritt eine sofortige Vasodilatation ein. Der muskuläre Tonus
nimmt dabei stark ab, sodaß die Gefäße den stufenweisen Drucksteigerungen
weit mehr als vorher nachgeben. Hierbei ist keine Spur von autoregulatori-
scher Vasokonstriktion mehr erkennbar. Zusatz von Extracalcium in steiger
der Konzentration zu der Tyrodelösung führt in den beiden rechten Bildab-
schnitten trotz weiterer Anwesenheit von Bay a 1040 zu einer Rückkehr des
Gefäßtonus und zu einer starken autoregulatorischen Vasokonstriktion selbst
auf niedrigen Druckstufen (Temperatur der Perfusionslösung 18 °C, pH 8, 3)

Hemmung arteriosklerotischer Gefäßprozesse durch prophylaktische Behandlung mit MgCl$_2$, KCl und organischen Ca^{++}-Antagonisten (Quantitative Studien mit Ca^{45} bei Ratten)

Von J. Janke, B. Hein, O. Pachinger, O. Leder und A. Fleckenstein
Physiologisches Institut der Universität Freiburg i. Br., Germany

Überdosierung von Vitamin D oder Dihydrotachysterol (DHT) führt insbesondere bei kombinierter Verabreichung mit NaH$_2$PO$_4$ bei Ratten nicht nur zu Myokard-Läsionen, sondern auch zu arteriosklerotischen Gefäßprozessen (1). Nach unseren vorausgegangenen Studien mit markiertem ^{45}Ca ist die Nekrotisierung durch eine Ca^{++}-Überladung der Myokardfasern bedingt und daher nach Schwere und Ausdehnung mit der ^{45}Ca-Inkorporation quantitativ korreliert (2, 3). Organische Ca^{++}-Antagonisten wie Verapamil, Substanz D 600 oder Prenylamin, die den transmembranären Ca^{++}-Influx in die Myokardfasern reduzieren, konnten daher die Nekrosebildung verhüten. Ähnlich wirkten K^{+}- und Mg^{++}-Salze als natürliche Ca^{++}-Antagonisten. Ganz analoge Verhältnisse wurden nunmehr an den Gefäßen gefunden: Ratten wurden 10 Tage lang peroral täglich mit 2 mal 10 mM/kg NaH$_2$PO$_4$ und 7 Tage lang peroral täglich mit 1 mal 0, 5 mg/kg DHT vorbehandelt. Am elften Tag wurde intraperitoneal ^{45}Ca (10 µC/kg) injiziert und die ^{45}Ca-Inkorporation in Aorta und Arteria mesenterica sup. gemessen. Bei den unbehandelten Kontrollratten erreichte der ^{45}Ca-Gehalt in der Aortenwand innerhalb von 6 Stunden nach der ^{45}Ca-Injektion 70 % und in der Mesenterial-Arterie etwa 100 % (bezogen auf die Plasma-Aktivität = 100 %), während bei den vorbehandelten Tieren der ^{45}Ca-Gehalt der Aorta innerhalb der gleichen Beobachtungszeit auf 5000 - 8000 %, d.h. auf das 70-120-fache und der Mesenterial-Arterie auf 4000 - 5000 %, d. h. auf das 40-50-fache der Norm anstieg. Histologisch ließen sich dabei intensive Media-Verkalkungen vom MÖNCKEBERG-Typ nachweisen. Gleichzeitig mit DHT und NaH$_2$PO$_4$ verabreichtes MgCl$_2$ bzw. KCl (jeweils 2 mal 7, 5 mM/kg täglich peroral) konnte die excessive ^{45}Ca-Inkorporation in die Gefäßmuskulatur der Arterien sowie die histologisch erfaßbaren Verkalkungen (ebenso wie die Myokardnekrosen) mehr oder weniger vollständig verhüten (vgl. Abb.). Organische Ca^{++}-Antagonisten (Verapamil, Prenylamin) wirkten gleichsinnig.

Literatur

1) SELYE, H.: Amer. Heart J. 55, 805-809 (1958).
2) FLECKENSTEIN, A.: Myokardstoffwechsel und Nekrose. In: Herzinfarkt und Schock, L. Heilmeyer und H. J. Holtmeier (Ed.). Stuttgart: Thieme 1968, pp 94-109.
3) JANKE, J., FLECKENSTEIN, A., JAEDICKE, W.: Pflügers Arch. ges. Physiol. 316, R10 (1960).
4) --- Pflügers Arch. ges. Physiol. 319, R8, R9 (1970).

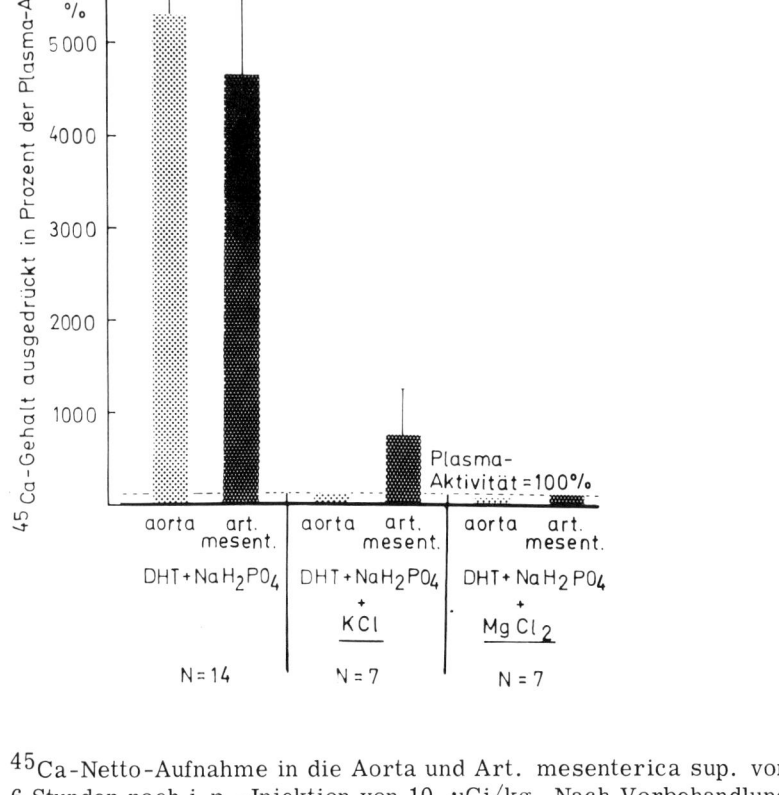

^{45}Ca-Netto-Aufnahme in die Aorta und Art. mesenterica sup. von Ratten
6 Stunden nach i. p. -Injektion von 10 µCi/kg. Nach Vorbehandlung der Ratten
mit NaH$_2$PO$_4$ (10 Tage 2 x 10 mM/kg p. o.) und Dihydrotachysterol (DHT)
(7 Tage 1 x 0,5 mg/kg p. o.) tritt innerhalb von 6 Stunden nach der ^{45}Ca-Ver-
abreichung eine excessive Überladung der arteriellen Gefäßwände mit Radio-
calcium ein. Durch prophylaktische Gaben von MgCl$_2$ oder KCl (10 Tage 2 x
7,5 mM/kg p. o.) kann diese pathologische Radiocalcium-Inkorporation mehr
oder wendiger vollständig verhindert werden, d. h. der Radiocalciumgehalt
der Aorta und Art. mesenterica sup. wird durch die MgCl$_2$- bzw. KCl-Pro-
phylaxe in den meisten Fällen in den Normbereich unbehandelter Kontroll-
tiere herabgedrückt, der etwa in Höhe der Plasmaaktivität (= 100 %) oder
etwas darunter liegt.

On the Control by Calcium of Rat Isolated Thoracic Aorta Responses to Noradrenaline and Barium

By J. C. Stoclet, F. Demesy, and M. Vendrely-Dolegeal
Laboratoire de Pharmacodynamie, Faculté de Pharmacie, Université Louis Pasteur, Strasbourg, France

The successive phases of tension development and relaxation were investigated on helical strips of thoracic aorta. Maximal isometric responses were elicited by noradrenaline (NA) and $BaCl_2$ in normal or Ca-free Krebs-bicarbonate solution (PSS). They were either potentiated by pretreatment in vivo with reserpine (Re, 5 mg/kg, i.p., 24 h) or depressed by addition of N-2'-0 dibutyryl-adenosine 3'5' monophosphate (cAMP, 1×10^{-7} M) or Papaverine (Pp., 0.5×10^{-5} M) in the PSS.

The shift from normal to Ca-free PSS first increases the relaxation rate (see fig.). It depressed then the tonic responses to NA and Ba^{2+} before the phasic response to NA. The magnitude of these depressive effects of Ca deprivation mainly depended of the number of responses elicited.

Pretreatment with Re potentiated the phasic and tonic responses and increased the relaxation rate in normal and Ca-free PSS.

Increase of PSS calcium concentration up to 10 mM during the tonic phase depressed weakly the response to NA but strongly potentiated the inhibitory activity of cAMP from exogenous or endogenous (Pp) origin.

These data suggest that extracellular Ca^{2+} controls not only the release of Ca^{2+} from intracellular binding sites during excitation-contraction coupling but the extrusion of Ca^{2+} from the cytoplasm as well. A tentative hypothesis is that the maximal tension developed in response to NA or Ba^{2+} may result from an equilibrium between the release and the extrusion of Ca^{2+}. Re would potentiate both processes and cAMP would increase the extrusion and (or) indirectly decrease the release, through membrane stabilisation. This hypothesis is also consistent with the observation that in Ca-free PSS the tonic component of the response elicited by NA is depressed before the phasic one.

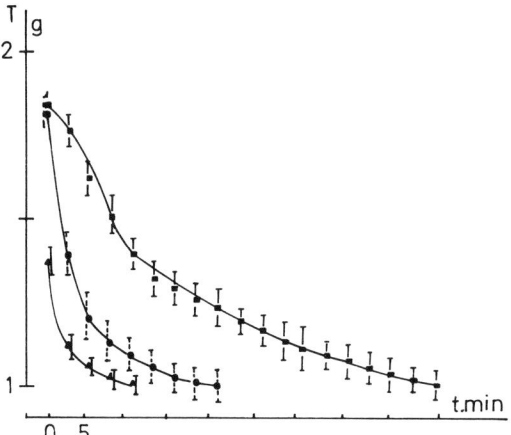

Effect of Calcium free solution on relaxation of aortic strips from the rat after response to noradrenaline:

■————■ = in normal Krebs-bicarbonate
●————● = first response in Calcium free Krebs-bibarbonate
▲————▲ = sixth response in Calcium free Krebs-bicarbonate.

Bath rinsed at time t = o.
Each value is the average of six experiments and vertical bars indicate fiducial limites (p : 0,05)

Further Studies on the Mediators of Functional Hyperaemia in Skeletal Muscle

By S. M. Hilton, O. Hudlická, and G. Vrbová
Department of Physiology, University of Birmingham Medical School, England

Increased osmolarity (MELLANDER et al., 1967), release of potassium (KJELLMER, 1965) and most recently release of inorganic phosphate (HILTON and VRBOVÁ, 1970) have been considered as the most likely candidates responsible for functional vasodilatation in contracting muscle. To evaluate the contribution of these factors, changes of osmolarity and of concentration of phosphate and potassium in the venous blood from muscle have been measured a) under conditions of graded muscular work where there is a relationship between increased blood flow and work performed, i. e. when work was increased by increasing the number of contracting units but not by increasing the load (ANREP and SAALFELD, 1935) and b) in slow and fast muscles since the slow muscle has little or no functional vasodilatation (HILTON et al., 1970).

In experiments on the cat gastrocnemius, if the work was increased by increasing the number of contracting muscle fibres, the release of potassium and phosphate was proportionate to the work performed. There was no relationship between the work performed and the changes in venous osmolarity. Increase in blood flow was very closely related to the release of inorganic phosphate. In the case of potassium, 20 % of the maximal release into venous blood was obtained in experiments in which the flow increased between 20 % and 70 % of the maximal value, and any further increase in potassium was associated with only a small increase in blood flow. No relationship was found between blood flow change and changes in venous osmolarity.

Functional hyperaemia is either very small or absent in the slow soleus muscle (HILTON et al., 1970). In fact such a hyperaemia was only seen in these muscles when the resting blood flow was not higher than 40 ml/100 g·min. Below this the smaller the resting flow the larger was the hyperaemia. The extent of the hyperaemia was not related to liberation of the potassium or to any increase in osmolarity. It occurred regularly when inorganic phosphate was released: The higher the release of phosphate, the greater was the hyperaemia.

In view of this evidence and that for the vasodilator activity of inorganic phosphate, this substance seems to be a good candidate as the agent initiating functional vasodilatation in skeletal muscle.

References

1) ANREP, G. V., von SAALFELD, E.: J. Physiol. 85, 375 (1935).
2) HILTON, S. M., JEFFRIES, M. G., VRBOVÁ, G.: J. Physiol. 206, 543 (1970).

3) HILTON, S. M., VRBOVA, G.: J. Physiol. 206, 29P (1970).
4) KJELLMER, I.: Acta physiol. scand. 63, 460 (1965).
5) MELLANDER, S., JOHANSSON, B., GRAY, S., JONSSON, O., LUND-VALL, J., LJUNG, B.: Angiologica 4, 310 (1967).

Effects of Hypoosmolality on Resistance to Blood Flow and upon Blood Pressure

By F. J. Haddy, J. B. Scott, D. K. Anderson, R. A. Brace, and D. P. Radawski
Michigan State University, E. Lansing, Michigan, USA

The effects of hyperosmolality on the resistance to blood flow through intact vascular beds and upon blood pressure are now well established. The effects of hypoosmolality are less well established because it is difficult to produce an isolated reduction in plasma sodium concentration and hence osmolality in blood perfused vascular beds and in all of the blood of the intact animal. In early studies we reduced the osmolality of the blood entering the canine forelimb and kidney (Amer. J. Cardiol. 8, 533 (1961); Amer. J. Physiol. 217, 1216 (1969); Amer. J. Physiol. 220, 384 (1971)) by rapid intra-arterial infusion of hypoosmotic solutions of sodium chloride, dextrose and urea and compared the resistance responses with those seen during equally rapid infusions of isosmotic solutions of the same agents. Resistance was higher during infusion of the hypoosmotic solutions, particularly in kidney, and additional infusions while measuring lymphatic vessel pressure, perfusing with cell free fluids and administering papaverine suggested that, in the kidney, the difference results, to a large extent, from active vasoconstriction subsequent to osmotic shift of water into the vascular smooth muscle cells. In the intact dog, rapid intravenous infusion of Ringer's solution low in sodium chloride content raised blood pressure while an equally rapid infusion of an isosmotic Ringer's solution was without effect (Amer. J. Physiol. 218, 234 1970)). We have recently developed non-dilutional methods for reducing the sodium concentration and hence the osmolality of the blood perfusing the isolated canine gracilis muscle and of all of the blood of the intact dog. A small dialyzer is interposed in the arterial supply of the gracilis muscle and the blood is dialyzed against a Ringer's solution low in sodium chloride content. This produces, within seconds, a large rise in the resistance to blood flow. All of the blood of the intact animal is rendered hyposmotic by administering a large dose of furosemide and replacing the lost urine with a hyposmotic Ringer's solution in which dextrose substitutes for sodium chloride. In this case the animal's own kidney serves as the exchange surface and an isolated fall in plasma sodium chloride concentration results within an hour, amoun-

ting to 9 mEq/l (plasma osmolality decreases 16 mOsm/kg). However, relative to suitable control animals, this degree of hypoosmolality is without
effect on blood pressure.

Studies in isolated tissues (BRADING and SETEKLEIV, J. Physiol. 195, 107
(1968); TOMITA, J. Physiol. 83, 450 (1966); GRUNDFEST, Fed. Proc. 26,
1613 (1967); JOHANSSON and JONSSON, Acta physiol. scand. 72, 456 (1968);
MELLANDER, JOHANSSON, GRAY, JONSSON, LUNDVALL and LJUNG, Angiologica 4, 310 (1967)) suggest that the active constriction produced by hypoosmolality results from increase in cell water, decrease in $[K^+]_i$, decrease
in membrane potential, and consequently rise in smooth muscle tension. Part
of the rise in resistance might result from passive vasoconstriction, subsequent to endothelial cell swelling, and increase in blood viscosity, subsequent
to red cell swelling.

Discussion

The discussion of the paper of ANDERSON, by HORN, JOHANSSON, FLEK-KENSTEIN concerned the mechanism of the decrease in resistance produced by an increase in plasma potassium concentration. Studies in a capillary viscometer indicate that blood viscosity does not decrease. Microelectrode studies in nerve and Purkinji fibers show that the membrane potential increases as the concentration of potassium in the bath is raised to 8 meq/1 (this is not what would be predicted from the Nernst equation). Perhaps hyperpolarization also occurs in vascular smooth muscle, therby accounting for the vasodilation

Discussion of the paper by GRAHAM concerned the origin of the spontaneous electrical activity seen in calcium-free solutions containing EDTA at 0-10 $^{\circ}$C. The pacemaker and spike itself were considered and the view was expressed that, regardless of origin, it appeared to be a regenerative phenomenon. The preparation was also considered. Determining the absolute temperature in the sucrose gap can be a problem.

KEATINGE discussing the paper by FLECKENSTEIN and GRÜN said he suspects the speakers are right in concluding that some of the agents studied act by blocking calcium entry into cells but warned against applying the hypothesis too generally. The speakers indicated that they have documented an inhibition of calcium entry in heart muscle but not in vascular smooth muscle. It is conceivable that the agents also inhibit sodium flux in less specialized membranes. Other areas considered were the molar potencies of the agents and their effects after depolarization with potassium.

Paper by GODFRAIND. Upon questioning, the speaker indicated that calcium content decreased in the presence of adrenaline but not in the presence of potassium. He also agreed that the methods do not distinguish between exchange in connective tissue and exchange in smooth muscle. One discussant indicated that norepinephrine slows down calcium exchange.

Paper by STOCLET. FLECKENSTEIN reviewed the two roles of calcium, namely membrane stabilization and excitation contraction coupling, in relation to the findings. Other discussion (BIAMINO, HEUSSLER) concerned the method of rendering the preparation calcium free.

Paper by HUDLICKA. DIETMANN and RODBARD asked: Is the rise in venous blood phosphate level sufficient to account for the vasodilation? Is it possible that the dilation results from increased concentrations of adenine nucleotides and adenosine and that the phosphates simply reflect this? The speaker pointed out that the concentration of phosphates surely is higher in extracellular fluid than in the venous effluent since capillary permeability to the phosphates is lower than to other small molecules, potassium for example. Adenosine could also act. One discussant pointed out that he failed to find increased levels of phosphates in the venous effluent but did find adenosine. Other discussion concerned the roles of metabolism and electrical activity in determining the phosphate levels.

Paper by HADDY. Discussion (RODBARD, STEVENS, FLOHR, JOHANSSON) concerned the mechanism of the increase in resistance seen during perfusion of vascular beds with blood having a low plasma osmolality. In addition to those mechanism suggested in the paper, rise in tissue pressure and effects on actomyosin were considered. The role of osmolality in autoregulation of renal blood flow and in exercise hyperemia were also discussed. It was pointed out that renal venous plasma osmolality does not change during renal autoregulation but that the osmolality of the venous effluent increases during exercise hyperemia.

Chairmen: F. J. Haddy, H. Rieckert

Druckabfall und Strömungswiderstand in den Arterien des Gehirn- und Extremitätenkreislaufes

Von E. Kanzow, J. Jansen und D. Dieckhoff
Physiologisches Institut der Universität Göttingen, Germany

Aus Arbeiten von FORBES, FOG, MCHEDLISHVILI u. a. ist bekannt, daß jede Zu- oder Abnahme des cerebrovasculären Widerstandes von entsprechenden Änderungen im Durchmesser der Piaarterien begleitet wird. An Arterien der hinteren Extremität fanden FLEISCH und HILTON eine Zunahme des Durchmessers während der Arbeitshyperämie. In welchem Ausmaß grössere und kleinere Arterien im Hirnkreislauf und in einem Haut-Muskel-Kreislaufgebiet zum Strömungswiderstand und seinen Veränderungen beitragen, sollten die Untersuchungen klären, deren bisherige Ergebnisse im folgenden zusammengefaßt werden.

Die Experimente wurden an Katzen in Chloralose-Urethan-Narkose ausgeführt. Gemessen wurden die Drucke in der Aorta (im Bereich des Aortenbogens), im rechten Vorhof und in einer peripheren Arterie des Hirnkreislaufs (A. cerebri media oder deren Äste) oder der hinteren Extremität (Interdigitalarterie oder A. femoralis). Alle Drucke wurden durch Katheter oder Mikrokatheter abgenommen und über elektrische Druckwandler fortlaufend aufgezeichnet. Nur in Gefäßen unter 250 μ Außendurchmesser und im Anfangsteil der A. cerebri media wurde der Druck nach Punktion mit angeschliffenen Glaskapillaren mit einem Gegendruckverfahren ermittelt.

Als Maß für den Anteil der Gefäße zwischen Aorta und punktierter bzw. katheterisierter peripheren Arterie am Gesamtwiderstand wurde der Druckabfall in dieser Gefäßstrecke in Prozent der Druckdifferenz Aorta - rechter Vorhof verwendet.

Der Druckabfall von der Aorta bis zu einigen Stellen des Hirnkreislaufes ist in Abb. 1 (oberer Teil) dargestellt. Jeder Punkt dieser Darstellung ist das Mittel mehrerer Messungen am gleichen Tier. Von der Aorta bis zur Carotis-Gabel (Druckmessung durch Katheter in der A. lingualis), d. h. auf einer Strecke von etwa 8-10 cm fällt der Druck bereits um etwa 5 % ab. Am Ursprung der A. cerebri media ist der Druck (gemessen mit Glaskanüle und Gegendruckverfahren) im Mittel um 24,7 % vermindert. Etwa der gleiche Wert ergibt sich bei Messungen mit Mikrokatheter in einem der großen Äste der A. cerebri media. In den kleineren Gefäßen fällt der Druck dann weiter ab. Bis zu den kleinsten punktierten Gefäßen (30-40 μ $\varnothing$) fällt der Druck um etwa 41 % ab. Die große Streuung der Meßpunkte ist Folge anatomischer und funktioneller Unterschiede zwischen den einzelnen Tieren sowie der unterschiedlichen Gefäßlängen bis zu den Meßpunkten in den einzelnen Experimenten.

Mit Unterstützung der Deutschen Forschungsgemeinschaft.

In den Arterien der hinteren Extremitäten (Abb. 1, unten rechts) wurde ein Druckabfall von 15-30 % (im Mittel 22,5 %) bis zur Pfote gefunden bei Messungen durch Mikrokatheter in Digital-Arterien (300-400 μ Ø). Ein deutlicher Druckabfall zeigte sich dabei in einigen Fällen bereits zwischen Aortenbogen und A. iliaca. In der A. femoralis war der Druck um etwa 10 % vermindert.

Aus beiden Diagrammen ergibt sich, daß nicht nur die kleineren sondern auch größere Arterien wesentlicher Teil der Strömungswiderstände im Hirn- und Extremitätenkreislauf sind.

Um festzustellen, in welchem Umfang der Strömungswiderstand in den Arterien auch zu Änderungen des Gesamtwiderstandes beiträgt, wurde am Hirnkreislauf der Druckabfall während CO_2-Hyperämie, an der Extremität der Einfluß von Papaverin und an beiden Kreislaufgebieten der Einfluß von Noradrenalin und Angiotensin untersucht.

Das Ergebnis dieser Experimente läßt sich dahingehend zusammenfassen, daß fast ausnahmslos der Druckabfall in den Arterien nach einer Anpassungsphase wieder dem Ausgangswert entspricht. Nur während Papaverin-Infusion wurde an den Extremitätenarterien eine deutliche Zunahme des Druckabfalls bei gleichzeitiger Verminderung des Aortendrucks beobachtet.

Um einen direkten pharmakologischen Einfluß auf die größeren Gefäße auszuschließen, wurde in weiteren Experimenten der Druckabfall zwischen Aorta und A. cerebri media untersucht bei Erhöhung des cerebrovasculären Widerstandes während zeitweiliger Steigerung des Aortendrucks (Autoregulation) durch Abklemmung der thorakalen Aorta mit Hilfe einer implantierten Schlinge. Abb. 1 zeigt unten links den Druckabfall etwa 1 Minute nach Beginn der Drucksteigerung bezogen auf den Ausgangswert. Auch hier bleibt in den meisten Fällen der Druckabfall nahezu unverändert. Wäre die Änderung des Gesamtwiderstandes auf letzteres beschränkt, müßte der relative Druckabfall fast in gleichem Ausmaß vermindert sein wie der Druckanstieg bzw. wie die Widerstandszunahme, in den vorliegenden Experimenten um 10-50 %. Da Verminderungen des Druckabfalls in dieser Größe nur in wenigen Fällen auftreten, nehmen an der Widerstandsänderung der Autoregulation sowohl die Gefäße, die die A. cerebri media versorgen, als auch die Piaarterien und präkapillaren Widerstandsgefäße teil.

Die hier dargestellten Ergebnisse lassen erkennen, daß größere und kleinere Arterien im Hirnkreislauf (und wahrscheinlich auch in anderen Kreislaufgebieten) so zusammenwirken, daß die Verteilung des Gesamtwiderstandes auf die einzelnen Abschnitte der Strombahn bei Widerstandsänderungen auf nahezu gleiche Werte eingestellt wird. Der Mechanismus, der dieses Zusammenwirken garantiert, ist möglicherweise eine Fortleitung von distalen Gefäßabschnitten nach proximal innerhalb der Gefäßwand, wie sie von FLEISCH und von HILTON in Arterien der hinteren Extremität nachgewiesen wurde.

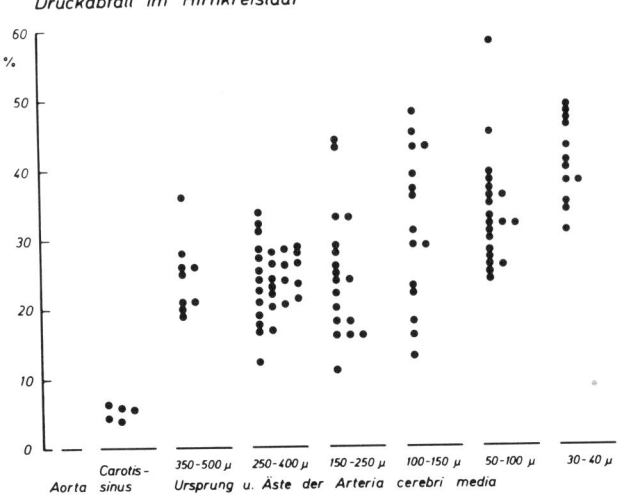

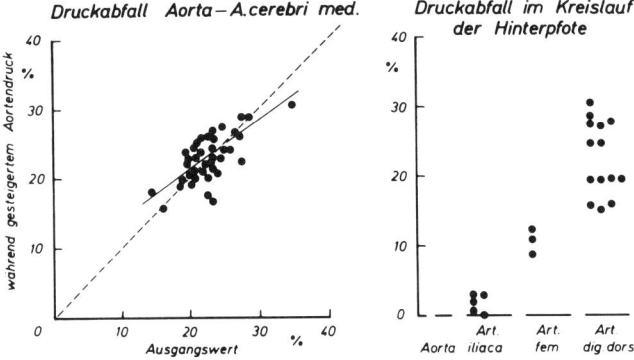

Abb. 1 oben: Druckabfall in den Arterien des Hirnkreislaufs in Prozent des Aortendrucks;

unten links: Druckabfall von der Aorta bis zur A. cerebri media (in Prozent des Aortendrucks) während erhöhtem Aortendruck und cerebrovasculärem Widerstand bezogen auf den Ausgangswert;

unten rechts: Druckabfall (in Prozent des Aortendrucks) im Kreislauf der hinteren Extremität.

Literatur

1) DIECKHOFF, D., KANZOW, E.: Über die Lokalisation des Strömungs-
   widerstandes im Hirnkreislauf. Pflügers Arch. 310, 75-85 (1969).
2) FLEISCH, A.: Les reflexes nutritifs ascendents producteur de dilatation
   arterielle. Arch. int. Physiol. 41, 141 (1935).
3) FOG, M.: Cerebral circulation. The reaction of the pial arteries to a fall
   in blood pressure. Arch. Neurol. Psychiat. (Chic.) 37, 351-364 (1937).
4) FORBES, H.S.: The cerebral circulation. Observation and measurement
   of pial vessels. Arch. Neurol. Psychiat. (Chic.) 19, 751 (1928).
5) HILTON, S.M.: A peripheral arterial conducting mechanism underlying
   dilatation of the femoral artery and concerned in functional vasodilatation in skeletal muscle. J. Physiol. 149, 93-111 (1959).
6) KANZOW, E., DIECKHOFF, D., HOLZGRAEFE, H.: Pressure drop in
   cerebral arteries at changes of the cerebrovascular resistance. In: Brain
   and blood flow, R.W.R. Russel (Ed.). London: 1971.
7) MCHEDLISHVILI, G.I., BARAMIDZE, D.G., NIKOLAISHVILI, L.S.:
   Functional behaviour of pial and cortical arteries in conditions of increased metabolic demand from the cerebral cortex. Nature 213, 506-507
   (1967).

The Effect of Varying Extracellular K+-, Mg++- and Ca++ on the Diameter of Pial Arterioles

By U. Knabe and E. Betz
Physiologisches Institut (I) der Universität, Tübingen, Germany

By means of microsurgical instruments small pial arteries and arterioles
were exposed. The extravascular space of these vessels was perfused with
mock CSF. For perfusion we used micro-pipettes.

In 11 cats we tested the effects of K^+ and Mg^{++} in non-physiological concentrations, the pH of the extracellular fluid beeing 7.23. This value was found
to be the normal value in cats. In 8 cats the effect of calcium deficiency on
the constrictory reactions of extravascular alkalosis was investigated. The
vascular reactions were recorded by microphotography.

K^+-ions: In 5 cats different parts of the pial vessels were studied
a) during application of K^+-free artificial CSF
b) in CSF with increased K^+ concentrations.
The H^+ was kept constant at a pH of 7.23. K^+-free artificial CSF caused in

4 of 5 cats vasoconstriction of the small pial arteries and arterioles. Increasing K^+-concentrations up to 13.12 mM/l (the 8 fold value of the normal CSF potassium) caused pial vascular dilatations. The degree of dilatation was dependent on the concentration. High concentrations caused more distinct dilatations (table 1).

Increasing K^+-concentration:

K^+-concentration mM/l	dilatation	constriction	no reaction
6.56	5 cats	1 cats	1 cats
13.12	3 "	- "	1 "
26.24	6 "	- "	- "
52.48	- "	- "	2 "
0	1 "	4 "	- "

The CSF pH was in all tests 7.23.

table 1

Mg^{++}-ions: In 6 cats the reactions of the pial vessels were recorded during increases in the extravascular Mg^{++} concentrations. The pH was kept constant at 7.23. We observed constrictory, dilatory and biphasic reactions (table 2).

Increasing Mg^{++}-concentration:

Mg^{++}-concentration mM/l	dilatation	constriction	no reaction
2.94	5 cats	5 cats	2 cats
5.88	3 "	1 "	- "
11.76	- "	- "	5 "

The CSF pH was in all tests 7.23.

table 2

Ca^{++}-ions: In 6 cats the pial vascular reactions were observed in alkaline Ca^{++}-free CSF. In two cats we added to the mock CSF 1 mM/l EDTA. The effects were recorded at pH-values of 7.8, 8.0, 8.1 and 8.2.

We observed three different types of reactions
1) During extravascular alkalosis the vessels constricted in the presence of Ca^{++}. In the absence of Ca^{++} the vessels did not constrict despite the alkalosis.
2) Some vessels showed only moderate constriction during alkalosis in the presence of Ca^{++} but a weak dilatation in Ca^{++}-free CSF.
3) In animals with severe respiratory acidosis no reaction was observed when the extravascular space was perfused with Ca^{++} containing alkaline CSF. The lack of Ca^{++}, however, caused strong dilatation.

Conclusion: The pial vessels are not able to develop constriction if the Ca^{++} concentration is reduced in the extravascular space. Ca^{++}-ions are one but not the only condition for developing pial vasoconstriction (table 3).

pH	with Ca^{++}: constriction without Ca^{++}: no react.	with Ca^{++}: constriction without Ca^{++}: dilatation	with Ca^{++}: no react. without Ca^{++}: dilatat.
7.80	4 cats	6 cats	2 cats
8.00	- "	1 "	- "
8.10	1 "	- "	- "
8.20	2 "	- "	1 "
+ EDTA:			
7.80	- "	1 "	1 "
8.00	2 "	- "	- "
8.20	- "	2 "	1 "
	9 cats	10 cats	5 cats

table 3

Recent experiments have demonstrated that an increase of the extravascular H^+-concentration causes dilatation and a decrease constriction of the pial arteries and arterioles. The question remains still unsolved whether H^+ changes act directly or via an interposed ionic mechanism upon the contractile structures of the pial vascular smooth muscles.

The Effect of Perivascular Microinjection of Norepinephrine and Epinephrine on the Diameter of Pia Arteries and Arterioles

By M. Wahl, W. Kuschinsky, O. Bosse, J. Olesen, N. A. Lassen, D. Ingvar and K. Thurau
Department of Physiology of the University of München; Department of Clinical Physiology Bispebjerg Hospital Copenhagen; Department of Clinical Neurophysiology of the University of Lund

The significance of catecholamines in the regulation of cerebral vessel resistance to blood flow was investigated by means of the perivascular space of pia arteries and arterioles of the cat. The local perivascular application of catecholamines is particularly suitable for this investigation, since vascular reactions can be detected independently of any changes in systemic hemodynamics, metabolism and acid base status. The method also excludes the modifying effect of the blood brain barrier on intravascularly applied substances The diameter of the vessels was measured from microphotographs taken of the cerebral surface. The catecholamines (norepinephrine and epinephrine 0.1, 1, 10, 100, 500, 1000 $\mu g/ml$) were solved in mock spinal fluid (11.5 mEq/l HCO_3^-) after the latter was found to have little or no vascular effect.

Both substances revealed a constrictory effect, which was localized in the vessel segment of which the perivascular space was punctured. The concentration-response curves obtained for both substances after subtraction of the bicarbonate effect are shown in fig. 1. When norepinephrine and epinephrine were applied in solutions whose pH was such as to cause constriction ($HCO_3^- = 22$ mEq/l) or dilatation ($HCO_3^- = 0$ mEq/l), the effect of pH was found to dominate. The catecholamine effect is strongly reduced in the alkaline and moderately decreased in the acidic solvent.

The results indicate that cerebral arteries and arterioles are sensitive to perivascularly secreted sympathetic transmitter substances. This is consistent with a sympathetic control of cerebral blood flow as indicated also by the results of JAMES et al. (Circulat. Res. 25, 77 (1969)) and D' ALECYAND FEIGL (Fed. Proc. 29, 520 (1970)).

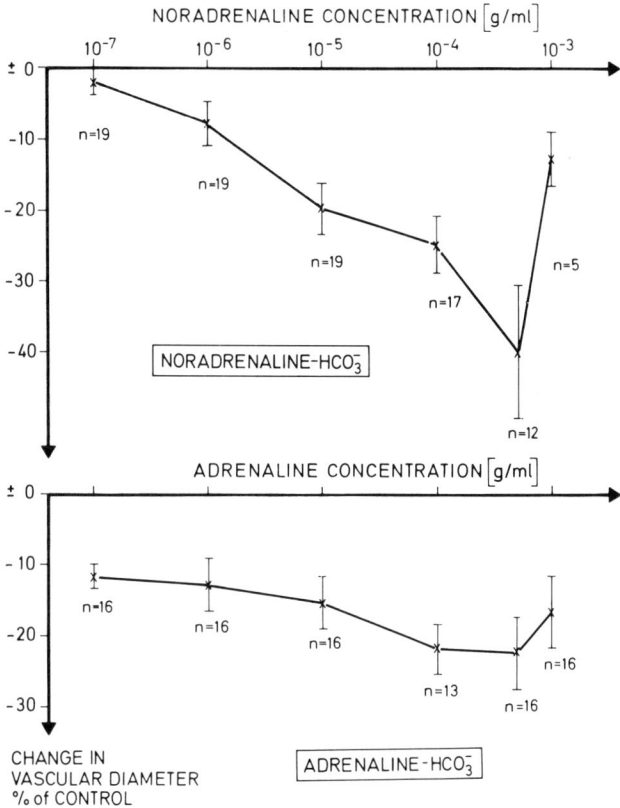

The concentration response curves of noradrenaline (upper panel) and adrenaline (lower panel). The curves demonstrate the effect due to catecholamines alone, which was obtained after subtraction of the effect due to the bicarbonate solvent.

α- and β-adrenergic Activity of Catecholamines in the Cerebral Vascular Bed

By G. Oberdörster, R. Lang, and R. Zimmer
Institut für normale und pathologische Physiology der Universität Köln, Germany

The effects of epinephrine (E) and norepinephrine (NOR) on cerebral blood flow (CBF) have often been studied by many investigators since the early experiments of WIGGERS (1907). Due to the various methods the results differed considerably. Recently the effects of isoprenaline (ISO) were investigated by LAUBIE et al. (1967) and JAMES et al. (1970).

The purpose of our studies was to show the effects of these three amines on cerebral vascular resistance (CVR) of the isolated perfused brain of the dog. Furthermore we wanted to study whether these effects could be increased, prevented or reversed by specific α - and β-blockade.

In 14 experiments canine brains (average weight 58 g) were perfused by donor dogs using the method described elsewhere (ZIMMER et al., 1971). The internal perfusion pressure in the circle of WILLIS was measured via the basilar artery; the external perfusion pressure was measured extracranially in the internal maxillary artery. Thus we were able to calculate not only changes in CVR but also changes of the resistance of extracerebral arteries leading to the brain.

This experimental technique allowed us to monitor continuously and quantitatively total cerebral blood flow without being influenced by changes of vascular resistance of different extracranial tissues. Mean cerebral oxygen consumption was 3.4 ± 0.4 ml/100 g $\cdot$ min.

Five applications were given of each dose of the three amines. The doses ranging from 0.001 μg - 10.0 μg were applicated intraarterially within 30 seconds in a volume of 0.1 ml.

The obtained dose response curves show E and NOR to have no statistically significant differences concerning their constrictor effects on the cerebral vasculature (table 1). Both amines lead to a dose dependent increase of CVR. ISO shows a dose dependent decrease of CVR.

Vascular resistance of the extracerebral arteries changes differently: A significant dilation (6 %) is seen after 0.01 μg E, while higher doses constrict these arteries. NOR leads always to a constriction of these vessels. The dilating potency of ISO on these extracerebral arteries is smaller than on cerebral vasculature (table 1). Due to these changes of vascular resistance CBF is decreased both by E and NOR in a dose dependent fashion. However, the decrease of CBF after 0.01 μg E is not statistically significant, that of 0.01 μg NOR is statistically significant. In the case of E this may be explained by dilation of the extracerebral arteries, while NOR leads to constriction of both the extracerebral and cerebral vessels. ISO shows a dose dependent increase of CBF.

After blockade of the α - and ß-receptors by phentolamine (2 mg) and propranolol (1 mg), respectively, equimolar doses (1 nmol) of the three amines
were given. The results are plotted in fig. 1.

table 1. Changes of cerebral vascular resistance (CVR, %) and of the resistance of extracerebral arteries (EA, %) following various doses of catecholamines.

Dose	Epinephrine		Norepinephrine		Isoprenaline	
g	CVR	EA	CVR	EA	CVR	EA
0.001	no effect		no effect		- 5	- 5
0.010	+ 7	- 6	+ 9	+ 3	- 15	- 8
0.100	+ 23	+ 6	+ 25	+ 6	- 32	- 13
1.000	+ 39	+ 17	+ 37	+ 14	- 48	- 19
10.000	+ 61	+ 113	+ 52	+ 91	- 51	- 25

The constrictor potency of NOR on CVR before (+ 15 %) and after ß-blockade
(+ 13 %) shows no significant difference. In contrast the constrictor potency
of E on CVR is significantly higher after ß-blockade (+ 26 %), while the dilating effect of ISO is completely prevented. The dilator potency of ISO on
CVR before (- 29 %) and after α -blockade (- 29 %) is unchanged. The constriction of the cerebral vessels after E and NOR is reversed into dilation
after α -blockade. The dilator potency of E (- 15 %) is higher than that of
NOR (- 9 %).

These statistically verified differences of the effects of 1 nmol of the three
catecholamines on the cerebral vessels can be summarized as follows:
1) The α -adrenergic potency of E : NOR : ISO is 1 : 0.5 : 0.
2) The ß -adrenergic potency of ISO : E : NOR is 1 : 0.5 : 0.3.

These results demonstrate the vasculature of the brain to have both α - and
ß-adrenergic receptors which can be stimulated by catecholamines in a similar way as in other vascular beds. These adrenergic receptors together with
those of extracerebral vessels may be important for the sympathetic regulation of cerebral blood flow.

References

1) LAUBIE, M., DROUILLAT, M.: Action de l'isoproterenol sur l'hemody-
    namique et le metabolisme cerebral du chien. Arch. int. Pharmacodyn.
    170, 93 (1967).
2) JAMES, I., XANALATOS, M.C., NASHAT, S.: The effects of changes in
    CO_2 tension and blood pressure on the response of cerebral metabolism

and blood flow to an isoprenaline infusion in the dog. CBF Symposium, London, 1970.

3) WIGGERS, C.J.: The innervation of the cerebral vessels as indicated by the action of drugs. Amer. J. Physiol. 20, 206 (1907).
4) ZIMMER, R., LANG, R., OBERDÖRSTER, G.: Post-ischemic reactive hyperemia of the isolated perfused brain of the dog. Pflügers Arch. (in press).

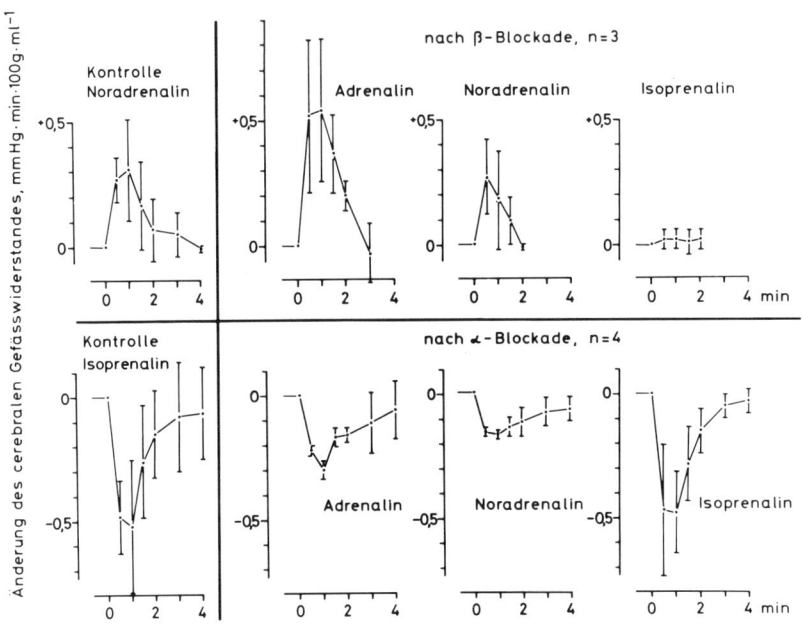

Changes of cerebral vascular resistance induced by 1nmol epinephrine, norepinephrine and isoprenaline before (Kontrolle) and after adrenergic blockade

Difference within the Cerebral Vascular Tree and between the Cerebral and Extracerebral Vessels in the Response Upon Vasoactive Compounds

By J. Olesen
Departments of Neurology and Clinical Physiology Bispebjerg Hospital,
Copenhagen, Denmark

Regional cerebral blood flow measurements in non anesthetized man are possible by means of the intra-arterial ^{133}Xe injection method. After percutaneous puncture of the common carotid artery a catheter is placed in the internal carotid artery and the tracer rapidly injected. The wash-out of radioactivity is followed by multiple collimated external scintillation detectors. The random experimental error is 4 to 5 % (1, 2).

It was found that the relationship between pCO_2 and CBF was best described as exponential in the pCO_2 range 25-60 mmHg. A pCO_2 change of 1 mmHg caused a 4 % change in CBF.

Injection of 10 mg papaverine into the internal carotid artery caused an increase of 93 % in cerebral blood flow. It was, however, observed that this increase could be obtained only in normal areas of the brain. In areas of cerebral infarction, hemorrhage or tumor the response was much smaller and even a paradoxical decrease of cerebral blood flow could sometimes be observed. The state of cerebral vasoparalysis which has previously been described for pCO_2 changes could thus also be demonstrated also for a pharmacological agent.

The most important factor causing vasoparalysis is probably acidosis. In a study of the vasoreactions to arterial blood pressure changes it was found that when the cerebral autoregulation was impaired it could usually be restored by hyperventilation (3). The concept of vasoparalysis is, however, of complex nature and also factors like tissue edema and intracranial pressure may play an important part in causing the abnormal flow changes.

Intracarotid infusion of histamine (7 patients) caused only a minor and non significant increase in CBF (average of all determinations: Rest 39.0 ml/ 100 g/min, histamine infusion 40.5 ml/100 g/min). In two patients lumbar CSF pressure was measured concomitantly with the CBF during infusion first of histamine then of papaverine. The CSF pressure increased with both substances, but CBF increase with papaverine was 100 % whereas no increase took place with histamine.

The rapidity of the pressure and flow changes seems to rule out pressure changes due to altered vascular permeability with formation of edema and rather indicates that they were caused by increased intravascular volume. If so then papaverine must be assumed to affect the resistance vessels whereas histamine appears to dilate vessels with negligible resistance.

Serotonin infused in the internal carotid artery in doses between 10 and 33

/ug/min caused no measurable change in CBF (mean of 8 determinations in 5 patients: Rest 44.5 ml/100 g/min, serotonin 43.5 ml/100 g/min). This is in agreement with animal experiments. A marked constriction of the extra-cranial part of the internal carotid artery was, however, seen with slightly higher doses of serotonin in animals (4). No obvious change in vessel diameter was seen in the large cerebral arteries (anterior, middle and posterior cerebral).

Adrenaline, noradrenaline and angiotensin were infused into the internal carotid artery as well as intravenously (5). No significant change in CBF was seen due to the internal carotid infusion. With intravenous infusion a significant increase in CBF took place. This was secondary to a marked increase in arterial blood pressure in some severely ill patients with impaired autoregulation and not to a pharmacological effect on the cerebral vessels. To test whether the substances had an effect on larger vessels but not on the small high resistance arterioles arteriography was repeated during infusion of noradrenaline and no caliber diminution was seen. Finally noradrenaline was administered during breathing of 8 % CO_2, a situation where the smaller vessels would be almost maximally dilated. No significant change in CBF took place. It was therefore concluded that not even on segments of the cerebral vessels did noradrenaline have a constrictor effect. Studies by WAHL et al. (6) this symposium clearly show that noradrenaline and adrenaline do cause vascular changes when applied to the pial surface of the vessels. This suggests importance of barrier mechanisms. Even these studies did, however, suggest less reactivity of cerebral than of other vessels.

On the contrary a strong constrictor effect of noradrenaline was seen in the extracerebral vessels. Even 5 % of the doses which were ineffective in creating changes of internal carotid circulation could severely constrict the external carotid vessels.

In conclusion the present data illustrate how vasoreactions in the brain are dependent on a normal milieu - especially pH. Different segments of the cerebral vascular tree may react differently to vasoactive compounds. When comparing the cerebral vessels to vessels of other organs the former are remarkably unresponsible to drugs. This may be due to blood - vessel wall barrier mechanisms but it may also be due to differences in the smooth muscle itself.

References

1) OLESEN, J., PAULSON, O.B., LASSEN, N.A.: Regional cerebral blood flow in man determined by the initial slope of the clearance of intraarterially injected [133]Xe. Stroke (in press).
2) SVEINSDOTTIR, E., THORLEV, P., RISBERG, J., INGVAR, D.H., LASSEN, N.A.: Calculation of regional cerebral blood flow (CBF). Initial-slope-index compared to height-over-total-area values. In: Brain and Blood Flow, Ross Russel, R.W. (Ed.). London: Pitman 1971.
3) PAULSON, O.B., OLESEN. J., STIG CHRISTENSEN, M.: Restoration of autoregulation of cerebral blood flow by hypocapnia. Neurology (in press).

4) DESHMEKH, V. D. , HARPER, A. M. : Effect of serotonin on cerebral blood flow and external carotid artery flow in the baboon. In: Brain and Blood Flow, Ross Russel, R. W. (Ed.). London: Pitman 1971.

5) OLESEN, J. : The influence of intra-carotid adrenaline, noradrenaline and angiotensin on the regional cerebral blood flow of man. Submitted to Neurology.

6) WAHL, M. , KUSCHINSKY, W. , BOSSE, O. , OLESEN, J. , LASSEN, N. A., INGVAR, D. H. , THURAU, K. : The effect of perivascular microinjection of norepinephrine and epinephrine on the diameter of pial arteries and arterioles. Satellite Symposium, 20-23 July 1971 Tübingen.

The Effects of Some Antimigraine Drugs on the Vascular Resistance in the External Carotid Bed of the Dog

By P. R. Saxena
Department of Pharmacology, Faculty of Medicine, Rotterdam, Netherlands

Migraine classically presents itself in two main phases. The initial pre-headache phase, sometimes absent, comprises of certain non-painful sensory symptoms usually attributed to intracerebral vasoconstriction (1, 2). The initial vasoconstriction is followed, during the headache period, by an excessive dilatation of the temporal, occipital and middle meningeal arteries (all of which are the branches of the external carotid artery). The latter phenomenom appears to be the chief source of pain in migraine (1). It is also known that the amplitude of pulsations and the blood flow in these extracranial vessels increases on the affected side during migraine headache (1, 3). The exact aetiology of the above changes is still uncertain. Nevertheless, certain recent findings implicate the involvement of one or more of the biogenic vasoneuroactive substances, such as 5-hydroxytryptamine (5-HT; serotonin), bradykinin and the similar peptides, histamine, substance P, acetylcholine and catecholamines, in the pathophysiology of the vascular pain (4, 5).

The present investigation was started with a view to study the effects of some recognized antimigraine drugs (ergotamine, methysergide, cyproheptadine) on the vascular resistance in the external carotid bed of the dog and the responses evoked by some vasoneuroactive substances on the above resistance.

A part of this work was performed in the Research Laboratoires of N. V. Organon, Oss, the Netherlands.

The experiments were performed in dogs anaesthetized with a mixture of 5 % α -chloralose and 25 % urethane (2 ml/kg). The animals were bilaterally vagotomized and given positive pressure ventilation. Arterial blood pressure was recorded from a cannulated femoral artery. Blood flow in the external carotid artery was recorded by placing an electromagnetic flow probe on the common carotid artery after ligation of its other branches. The resistance in the external carotid bed was calculated by dividing the arterial pressure in mmHg by the flow in ml/min; assuming the venous pressure to be zero. The effect of intravenous cumulative doses (1, 1, 2, 4, ... sequence) of three antimigraine drugs were studied on the external carotid bed resistance, and on the vascular responses elicited by intra-arterial administration (into cannu lated common carotid artery) of 5-HT (25 μg), histamine (2 μg), noradrenaline (2.5 μg) and bradykinin (0.5 μg). The response (Res) of their agonists was calculated from the formula: $Res = |1 - Ra/Rb|$ i.e. deviation from 1 of the ratio of resistance at their peak effect (Ra) and that just before their administration (Rb).

Ergotamine (1-32 μg/kg) and methysergide (0.02-0.64 mg/kg) caused an enormous decrease (respectively, 32-77 % and 12-59 % from control values) in the external carotid blood flow resulting from vasoconstriction leading to an increase (respectively, 53-736 % and 17-229 %) in the external carotid bed resistance. Cyproheptadine (0.125-2.0 mg/kg) had no significant effect on the above two parameters. Since there was no concomitant change in the blood pressure by ergotamine (up to 8 μg/kg) and methysergide, it seems likely that the vasoconstriction in the external carotid bed was selective.

Intra-arterial injection of 5-HT and noradrenaline elicited a vasoconstriction, while that of histamine and bradykinin caused a vasodilatation in the external carotid bed. In the same doses as employed above, ergotamine (-53 to -158 % from control) and methysergide (0 to -49 %), but not cyproheptadine, attenuated the response to 5-HT; most probably by altering the vascular tone in the above bed; see HADDY et al. (6). There was enhancement in the response to noradrenaline (+27 to +108 %) and histamine (+11 to +39 %) by methysergide; and to bradykinin (+31 to +81 %) by ergotamine. The last-mentioned effect could be due to the enormous change in the control vascular resistance. Cyproheptadine significantly reduced the histamine-effect (-24 to -50 %).

The results of the present study cast doubts on the assumption that the antimigraine drugs owe their therapeutic usefulness to their antiserotonin action. The selective vasoconstriction elicited mainly by ergotamine and to some extent by methysergide in the external carotid bed (which in dogs includes a part of intracerebral circulation) appears to be a more likely explanation for their therapeutic effectiveness since cyproheptadine, which lacks such an action but antagonises histamine-response, has come out inferior to methysergide in migraine trials (7, 8). It is likely that the noradrenaline-potentiating effect of methysergide, may also contribute to its therapeutic efficacy in migraine.

References

1) WOLFF, H. G. : Headache and other head pains. 2nd ed. Oxford: University Press 1963.
2) DUKES, H. T. , VIETH, R. G. : Neurology 14, 636-639 (1964).
3) ELKIND, A. H. , FRIEDMAN, A. P. , GROSSMAN, J. : Neurology 14, 24-30 (1964).
4) SICUTERI, F. : Clin. Stud. Headache 1, 6-45 (1967).
5) FRIEDMAN, A. P. : "Handbook of Clinical Neurology". Vol. 5, Vinken, P. J. and Bruyn, G. W. (Eds.). North Holland Publishing Co. 1968, pp. 37-44.
6) HADDY, F. J. et al. : Circulat. Res. 7, 123-130 (1959).
7) CURRAN, D. A. , LANCE, J. W. : J. Neurol. Neurosurg. Psychiat. 27, 463-469 (1964).
8) LANCE, J. W. : Proc. Aust. Ass. Neurol. 4, 81-84 (1966).

Reactive Changes of Cerebral Vascular Resistance at Different Transmural Pressure

By K. Held, W. Niedermayer, W. Gottstein and J. Schaefer
I. Medizinische Klinik der Universität Kiel, Germany

The series of experiments reported here have initially been started under clinical aspects. We were interested how cerebral blood flow (CBF) and metabolism are influenced by different forms of assisted circulation, which are used in clinical medicine. The emerging results may,however, also be of some relevance for the understanding of cerebral autoregulation.

The experiments have been performed in dogs, cerebral venous outflow obtained by the method of RAPELA and coworkers has been monitored continuously with an electromagnetic flowmeter.

We studied in the first set of experiments the effect of a non-pulsatile perfusion on CBF - a condition that occurs frequently in cardiac and vascular surgery during extracorporeal circulation. A Windkessel between the aortic arch and brain virtually flattened out all pulsations of pressure and flow. A pump behind the Windkessel restored a pulsatile perfusion that equaled normal conditions in pressure and flow.

The first slide demonstrates a typical observation: When perfusion is changed from non-pulsatile to a pulsatile one, venous flow rises temporarily, but returns to the initial level within short, followed by an increase of pressure.

Because driving pressure is delivered at a constant rate and volume by the pump the observed pressure-elevation in face of a constant flow is taken as an indication of an increased vascular resistance. Since other factors, well-known to influence CBF as pH, pO_2 and pCO_2 were kept constant, we concluded that the observed effect upon CVR was due to the change of pressure-configuration only.

If one applies the Bayliss-concept it appears reasonnable to assume that a pulsatile pressure has to be more effective in evoking a vascular reaction than a non-pulsatile or "mean" pressure. At this point we were even wondering whether such a mean pressure could evoke the typical autoregulatory response at all. We therefore tested cerebral circulation with both forms of perfusion pressure and obtained the following results: An effective autoregulation was observed in both instances, but on a lower flow level in pulsatile perfusion (lower curve). With other words: CVR again is higher at pulsatile pressure than in the absence of pulsations.

Here I have to admit that we weren't aware so far of two points, which have been discussed in the literature earlier and which are quite pertinent to our question.

The first is quite an old argument of German physiologists wether a pulsatile or a non-pulsatile pressure is more effective in propagating organ blood flow. SSENTJURIN in 1927 concluded from his studies - I quote in my translation - "the pulsating flow does not cause a vascular dilatation, but rather a certain constriction of the vessels".

And secondly the results of WEISS, who described in 1959 findings quite similar to ours in the regulation of blood flow in isolated kidneys.

The results of our experimental work and studies of the literature seemed to support our assumption, that the recurrent pulsations exert a stronger constrictory stimulus than an identical mean pressure. If this is true, then even stronger responses should be expected if pressure pulsations were multiplied. This was achieved with a method, which again is used in clinical medicine as a special means of assisted circulation. In this procedure a pump generates a pressure volume puls, that is directed against the aortic flow and triggered to fall into cardiac diastole. The method therefore is called diastolic arterial counterpulsation. The slide demonstrates the doubling of pressure pulsations per cycle. In these experiments mean arterial pressure remains essentially constant during counterpulsation after an initial shortlasting decline. CBF, however, reacts to the multiplied pulsation in the anticipated manner: It falls in the steady state to about 90 % of the initial value, because CVR rises by about 10 %. The extensive variations in CBF reduction depend mainly upon the amplitude of the second, diastolic pressure wave: Lower ones of about 20 mmHg are almost ineffective, whereas higher amplitudes of about 50 mm Hg increase CVR considerably.

So far we have demonstrated that by simply changing the form of perfusion pressure from a non-pulsatile over a normal pulsating to a doubled pulsatile one - keeping mean pressure quite constant - considerable variations of CVR were induced. We feel inclined to regard these reactive changes of CVR as an indication of vascular responses to the altered configuration of the transmural pressure - mediated perhaps by a myogenic mechanism.

The pressure induced changes of CVR apparently contain two components: One sensitive to constant pressure, that could be called the DC-component and another one reacting upon pressure changes, which then consequently had to be an AC-component. The latter should be dependent on the amplitude of pressure pulsations - which has indeed been shown in the last set of experiments. Furthermore it has to react upon the velocity of pressure-changes: The quotient of dp/dt, as indicated by the strong response to normal and doubled pulsations and tested further in another study where a square wave pressure puls was applied to the regional vascular area of the middle cerebral artery. Preliminary results of these experiments which are done together with SYMON, London, indicate a fast response of CVR, occuring within less than 1 second, that is induced by the rapid change of perfusion pressure.

In conclusion we have demonstrated definite changes of CVR of an identical mean pressure, which are induced by variations of the pressure profile only. The pressure induced changes of CVR are of very short latency as also demonstrated in renal and coronary circulation by WEISS and coworkers. In our view these results support the myogenic concept of cerebral autoregulation.

Diskussion

Vortrag KANZOW

BETZ: Haben Sie Unterschiede in der Papaverin-Wirkung während Hypercapnie und Hypocapnie gefunden, da ja doch bekannt ist, daß auch die Veränderung der H-Ionen-Konzentration einen Einfluß auf die cerebrale Gefäßweite besitzen?

KANZOW: Für Papaverin haben wir keine Experimente in Hypo- und Hypercapnie durchgeführt. Mit Noradrenalin sind schon Versuche gemacht worden, aber die Interpretation scheint sehr kompliziert zu sein. Wir wissen aber noch nichts Genaues darüber.

Vortrag KNABE

KUSCHINSKI: Wir haben, was die Wirkungen von H^+- und K^+-Ionen angeht, die gleichen Ergebnisse bei ähnlicher Methode bekommen. Ich kann noch einige quantitative Angaben hierzu machen. Trägt man in einem Koordinatensystem auf der Ordinate den Gefäßdurchmesser, auf der Abszisse die Bikarbonatkonzentration auf, so findet man bei Bikarbonatkonzentration Null eine Dilatation, bei 11 mEq/l nur geringe Reaktionen, bei 22 mEq/l deutliche Konstriktionen. Trägt man nun auf der Abszisse statt der Bikarbonatkonzentration die Kaliumkonzentration von 0-20 mEq/l bei einer Bikarbonatkonzentration von 11 mEq/l auf, so findet man eine gute Korrelation zwischen Gefäßwiderstand und Kalium.

LASSEN: Haben Sie irgendwelche Unterschiede zwischen großen und kleinen Arterien gefunden?

BETZ: Wir haben die Effekte bei großen Arterien nicht beobachten können, sondern nur bei Gefäßen eines Durchmessers von weniger als 300 /um. Das mag allerdings z. T. daran liegen, daß die Verengerung großer Gefäße nicht so gut erkennbar ist. Es scheint mir aber noch ein weiteres Problem zu bestehen, das die Veränderungen des osmotischen Druckes durch die Ionen-Konzentrationsänderungen direkt beim Gefäß betrifft. Wir haben bestehende osmotische Druckunterschiede mit Glucose ausgeglichen. Dabei stellten wir fest, daß die verursachten Wirkungen die Ergebnisse nicht wesentlich beeinflußten.

WAHL: Ich kann hier beipflichten, der Einfluß ist wirklich nicht sehr groß. Wir haben den osmotischen Druck durch Austausch von Kalium für Natrium konstant gehalten.

Vortrag WAHL

BETZ: Es erhebt sich natürlich die Frage nach der Interaktion der Ionen auf die Gefäßmuskeln. Wirken die pH-Veränderungen über Verschiebungen des K^+? Für die Potentialänderungen an der glatten Muskelzelle scheinen die Kaliumionen die wichtigeren zu sein, denn die durch H^+-Änderungen entstehenden Potentialdifferenzen sind im physiologischen Bereich sehr gering.

WAHL: Ja, das scheint tatsächlich der Fall zu sein. Kalium spielt eine große Rolle. Nur bei sehr niedrigen Bikarbonatkonzentrationen zwischen 0 und 5 mEq/l finden wir eine Dominanz der H-Ionenkonzentration, im anderen Bereich zwischen 5 und 22 mEq/l haben wir eine klare Dominanz der Kalium-

konzentrationen über die Bikarbonatkonzentration gefunden.
GREEN: Es bestehen ja doch sehr große Unterschiede zwischen Ihren Nor-
adrenalin-Ergebnissen und den Ergebnissen von Dr. Rapela, der bei einer
Sympathicus-Reizung keine Änderungen gefunden hat. Eine Erklärung kann
vielleicht dadurch gegeben werden, daß die Piagefäße anders als die intra-
cerebralen Gefäße reagieren, zum anderen kann es sein, daß es im Markbe-
reich kompensatorisch zu einer Erweiterung kommt, sodaß insgesamt die
Gehirndurchblutung konstant bleibt.
WAHL: Wir wissen, daß Piagefäße ganz ähnlich wie andere cerebrale Gefäße
reagieren, jedenfalls bei Änderungen der Konzentration von CO_2, H^+-Ionen
und Kalium. Ich kann mir die andersartigen Ergebnisse von Dr. Rapela nicht
erklären.
LEUSEN: Wir haben keine Erfahrungen über Kaliumwirkungen, aber ich kann
einige Bemerkungen zu Bikarbonatwirkungen in tieferen Strukturen machen,
und zwar im Bereich des Nucleus caudatus. Wir haben bei lokalen Bikarbo-
natinjektionen mit Wärmeleitelementen und Xenonclearance die Gehirndurchblutung gemessen, wobei wir bei den Durchblutungsmessungen mit Wärmeleitelementen sehr schöne, pH-korrelierte, lokale Änderungen gefunden haben, die allerdings bei den globalen Clearancemessungen nicht zutage traten.

Vortrag OBERDÖRSTER

BETZ: Wir haben sehr oft die gleiche Reaktion auf schnelle i. v. -Injektionen
von Noradrenalin und Adrenalin gefunden. Gleichzeitig mit der initialen
Durchblutungserhöhung kam es dabei zu CO_2 Auswaschungen aus dem Gewebe,
verbunden mit einer entsprechenden pH-Änderung, die dann wiederum eine
Gefäßkonstriktion herbeiführte. Wir haben aber auch manchmal Dilatationen
mit Anstieg des pCO_2 auf den Cortex gesehen und dabei etwas geringeren
Blutdruckanstieg; also eine Dilatation im Gegensatz zur normalerweise auftretenden Konstriktion. Gewöhnlich kam es nach der Konstriktion zu einer
reaktiven Gefäßerweiterung, d. h. wir haben auch biphasische Reaktionen beobachtet. Haben Sie etwas Ähnliches gesehen?
OBERDÖRSTER: Nein, wir haben nie biphasische Reaktionen gesehen.
HELD: Ich habe noch eine methodische Frage. Haben Sie bei diesem relativ
großen, präparativen Aufwand die Autoregulation geprüft?
OBERDÖRSTER: Ja, zwischen 60 und 100 mmHg war die Autoregulation vorhanden.
KANZOW: Meine Frage geht in der gleichen Richtung. Wir haben nämlich bei
wachen Tieren keine Durchblutungsminderung auf Infusion von Noradrenalin
gefunden, sondern eine konstante Durchblutung, die wir durch eine intakte
Autoregulation erklären. Ich bezweifle demnach doch noch, daß die Autoregulation bei Ihren Tieren intakt war, nämlich gerade durch eine nicht intakte
Autoregulation konnten Sie Ihre Effekte ja gut nachweisen. Haben Sie denn
einmal Angiotensin gegeben?
OBERDÖRSTER: Nein, wir haben Angiotensin noch nicht ausprobiert.

Vortrag OLESEN

FIESCHI: Wir haben bei akuten apoplektischen Insulten totale Vasoparalysen
mit einem Verlust der Autoregulation und Verlust der Reaktivität auf CO_2-

Änderungen beobachtet.

OLESEN: In leichteren Fällen bemerkten wir oft eine Diskrepanz zwischen Verlust der Autoregulation und intakter Reaktivität auf CO_2, doch in sehr schweren Fällen sind sowohl CO_2-Reaktivität als auch Autoregulation nicht mehr intakt.

Vortrag HELD

FLOHR: Nach Ihren Ausführungen ist das Auftreten der Autoregulation ein sehr kurz dauernder Prozeß, der etwa eine Pulswelle dauert, und das steht im Gegensatz zu unseren sonstigen Beobachtungen, wo bei den Änderungen des Blutdrucks die Autoregulation Sekunden, sogar Minuten in Anspruch nimmt.

HELD: Ich bin überzeugt, daß die Autoregulation bei Änderung des Blutdruck innerhalb Bruchteilen von Sekunden vor sich geht.

FLOHR: Ich kann Ihrer Schlußfolgerung nicht zustimmen, daß die Autoregulation eine Funktion des transmuralen Druckes ist. Wenn Sie den Fluß ändern, dann können Sie sehr leicht zu der metabolischen Theorie der Autoregulation kommen.

HELD: Ich bin überzeugt, daß es der transmurale Druck ist. Ich ändere nämlich nicht den Fluß, sondern die Druckkonfiguration, und ist für mich nicht einsehbar, weshalb dann noch ein metabolischer oder nervöser Effekt eine Rolle spielen soll. Wenn man das Bayliss-Konzept anwendet, kann man durch aus zu der Meinung kommen, daß es eben der transmurale Druck ist.

Chairmen: N. A. Lassen, J. A. Bevan

The Mechanism of Closure of Sheep and Human Umbilical Arteries

By M. R. Roach
Departments of Biophysics and Medicine, University of Western Ontario,
London 72, Ontario, Canada

Umbilical arteries are large arteries (diameter 6-8 mm in sheep and humans)
which must close after birth before the cord ruptures. We believe this closure is produced largely by longitudinal rather than circular muscles. These
muscles shorten and bulge into the lumen, effectively plugging it. Since the
volume of muscle must remain constant, the degree of shortening (25-30 %
in sheep) should equal the increase in cross-sectional area of the wall (26-29
% in same arteries). The segmentation seen in the human arteries, but not
in the sheep, is probably due to a different organization of the longitudinal
muscle, with the ends "lined up" in humans and overlapping in sheep. Calcu-
lations show that the force required to overcome the intraluminal pressure
is twice as great for circular as for longitudinal muscle. We found the stimulus for contraction of the longitudinal muscles in the human was a transient
stretch. Oxygen was required, but did not cause contraction by itself. Various drugs, gases, etc. had no effect on the longitudinal muscle but did stimulate the circular muscle. The coiling muscle for the human artery contracted
when the temperature dropped to $27 \pm 0.5\ ^{\circ}C$. This illustrates the importance
of understanding muscle geometry in designing experiments to test effective
stimuli for muscle contraction.

Supported by the Medical Research Council of Canada.

The Effect of Phenoxybenzamine and Propranolol on the Circulation of Pregnant Rats

By L. A. Debreczeni and L. Takács
II. Department of Medicine, Semmelweis Medical School Budapest, Hungary

The effect of phenoxybenzamine Phe : 10 mg/kg i. p. and propranolol Pp : 2 mg/kg i. v. pretreatments were studied on the cardiac output CO : Evans blue dilution and its organ fractions ^{86}Rb isotope indicator fractionation tech nique (SAPIRSTEIN, L. A., Amer. J. Physiol. 193, 161-168 (1958)) of dioe- strual and pregnant rats. The weight of the pregnant uterus varied between 10-50 g.

A change in the vascular tone was considered if flow and resistance changed simultaneously and inversely. CO/ml/min/100 g body weight of the pregnant controls compared to the dioestrual ones decreased. BP, TPR and the vascu lar tone of the coronary, renal and spleen arteries of the pregnants increa- sed.

In dioestrual rats CO did not change after pretreatments. Phe decreased TPR, Pp increased it. After Pp increased vascular tone of spleen, skin and carcass vessels were detected.

In pregnant rats Phe elevated CO and decreased BP and TPR. After Phe vas cular tone of coronary, renal, gut, skin, carcass vessels decreased. Nei- ther Phe nor Pp influenced the vascular tone of the pregnant uterus.

The pregnant uterine resistance of the groups did have a significant correla- tion with increasing uterine weight and this was not influenced by Phe or Pp (fig. 1). The slopes of regressions did not differ significantly.

In dioestrual rats no significant alpha sympathetic control of the vessels was observed.

In pregnant rats an intensive sympathetic tone of alpha type seems to control regional blood flow - also in the kidneys - with the exception of the uterus. Regarding the renal vascular effects of Phe the present findings contradict our previous results as under control conditions renal blood flow is not influ enced by alpha blocking agents (DEBRECZENI, L. A., TAKÁCS, L., Acta physiol. Acad. Sci. hung. 33, 169-183 (1968)).

After Pp increased vascular tone of spleen, skin and carcass vessels of the non-pregnant rats were detected. In contrast to this in pregnant animals Pp pretreatment increased only the vascular tone of the renal vessels.

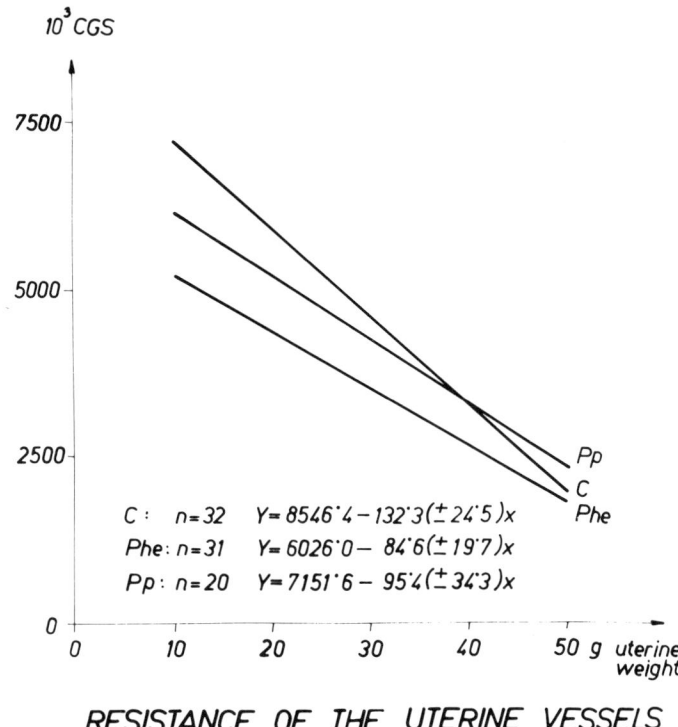

$10^3 CGS$

7500

5000

2500

C : n=32   Y= 8546'4 - 132'3(±24'5)x
Phe: n=31   Y= 6026'0 - 84'6(±19'7)x
Pp: n=20   Y= 7151'6 - 95'4(±34'3)x

Pp
C
Phe

0

0   10   20   30   40   50 g uterine weight

RESISTANCE OF THE UTERINE VESSELS

Zur Autoregulation der Nierendurchblutung

Von K. Held, W. Niedermayer, J. Schaefer, H. J. Schwarzkopf und Ch. Weiss
I. Medizinische Universitätsklinik Kiel und Physioolgisches Institut der
Universität Kiel, Lehrstuhl II, Germany

Zur Prüfung des Einflußes primärer Eigenschaften der glatten Muskelfaser
auf die Durchblutungsautoregulation der Niere haben wir an Hunden mit der
Technik der arteriellen Gegenpulsation künstliche, das arterielle Druckpro-
fil in bestimmten Phasen überlagernde Druckanstiege induziert.

Dabei zeigte sich, daß die hierdurch erzielbare Änderung des Strömungswi-
derstandes der Niere in 2 Phasen verläuft: Einer kurzdauernden, rasch einsetzenden (nach 0, 5-1 sec. beginnend) Komponente folgt eine verzögert auf-
tretende (5-10 sec. nach Druckanstieg), anhaltende, vom Ausmaß der Druck-
zunahme abhängige Komponente.

Die erhobenen Befunde werden unter Berücksichtigung der Untersuchungsergebnisse an der isolierten Niere und der Taenia coli des Meerschweinchens,
welche ebenfalls eine reaktive biphasische Änderung des Perfusionswider-
standes bzw. eine reaktive Spannungsentwicklung mit zwei "Resonanzstellen"
zeigen, diskutiert.

Diese druckinduzierten Änderungen der Nierenperfusion werden als Bestand-
teil der Autoregulation der Durchblutung des Organs diskutiert.

The Arterial Inflow to the Forearm of Normal Subjects and Hypertensive Patients Following Sudden Changes in Local Vascular Transmural Pressure

By P. H. Fentem, A. D. M. Greenfield, and J. Yates
Department of Physiology,The Medical School, University of Nottingham,
England

ARDILL, FENTEM and ISAAC (1969) showed that when the forearm was exposed to pressures below atmospheric, there was an initial increase in the
arterial inflow to the limb. In that study the observations were made on 6
healthy young men. The relation between the arterial inflow during the first
few seconds of suction and the decrease in external pressure was curvilinear
with the convexity towards the pressure axis. When the pressure around the
limb was 90 mmHg below atmospheric inflow was increased more than five-
fold.

Similar measurements have now been made on the change in arterial inflow t
the forearm of 8 hypertensive patients (mean B. P.: 130-177 mmHg) and the sar
number of normotensive subjects matched for age and sex. The results were
analyzed by determining at several different pressures the ratio between arte
rial inflow during the first few seconds of suction and the mean resting level
of arterial inflow through the forearm immediately before the suction. In the
hypertensive subjects the arterial inflow during suction was smaller than tha
observed in the normotensive subjects for suction pressures within the range
0 to -100 mmHg; For example in the normotensive subjects flow was increased fivefold at -80 mmHg whereas in hypertensive patients it was increased
only threefold. Regression lines drawn for the relationship between the logarithm of the ratio of inflow (during suction/before suction) and the suction
pressure were significantly different both as regards intercept and slope (ana
lysis of co-variance: slope $P < 0.001$ intercept $P < 0.001$). These results ap

This work was supported by a grant from the British Heart Foundation to
Professor A. D. M. Greenfield and Dr. P. H. Fentem.

pear to demonstrate that the arterioles which have a smaller calibre in established hypertension resist the dilating effect of sudden changes in transmural pressure more effectively than normal vessels.

Both groups of subjects were older than subjects investigated previously (ARDILL et al., 1969) so that in a further study a comparison was made between normotensive subjects of two age groups, women 20-25 and 55-90. In the group of older women the arterial inflow was smaller than that observed in the younger group at all pressures. However, the difference between these two groups was much smaller than that between the hypertensive and the normotensive subjects. The regression lines drawn for the relationship between the logarithm of the ratio of inflow were of almost identical slope but the intercepts were significantly different (analysis of co-variance; intercept P < 0.01).

Following the dilatation induced by a sudden increase in transmural pressure the resistance vessels respond by an increase in tone, that is there is reactive constriction. Measurements of forearm blood flow were made immediately after an exposure to a negative pressure of 80 mmHg lasting 8 seconds. In both the normotensive and hypertensive subjects the blood flow fell to 70 % of the value recorded before the suction. The lowest values were recorded 20 seconds after the end of the suction. The results were identical in the two groups.

From this study we concluded that the resistance vessels of the forearm of hypertensive subjects differed from those of normotensive subjects in their response to sudden changes in transmural pressure. The vessels did not dilate so dramatically as did apparently normal vessels. Age appeared to have a similar effect to the hypertensive state, though it was much smaller in this study. The reactive constriction which followed the period of suction was similar in both groups.

References

1) ARDILL, B. L., FENTEM, P. H., FINLAY, R. D., ISAAC, P.: J. Physiol. 203, 31-43 (1969).

Discussion

In the discussion Dr. ROACH was asked whether the longitudinal muscle is spontaneously active. She answered that it contracts very quickly after delivery. So it can only be studied if the cord is cross-clamped in order to prevent the contraction. This is not the same mechanism as that during critical closure.
NIEDERMAYER mentioned that he saw pressure-induced changes of kidney blood flow not only in innervated kidney but also during local anesthesia of the kidney. When asked by RODBARD how much in his results the capillaries influenced his results he answered that he considered this, but the change of flow was too great to be explained by dilatation of the capillaries.

Chairmen: M.R. Roach, K. Golenhofen

Evidence of Regulation of the Caliber of the Cerebral Vessels by the Extra-Cellular pH. Studies on the CBF Adaptation During Chronic Respiratory Acidosis

By A. Agnoli, N. Battistini, M. Casacchia, M. Nardini and C. Fieschi with the technical aid of S. Passero
Clinica delle Malattie Nervose e Mentali della Universita de Roma, e Clinica Nervose e Mentali de Siena, Italia

The regulation of the caliber of the cerebral vessels has been related to the pH of the fluid of the smooth fibers of the vessel walls.

This hypothesis, deriving from observation on the relationship between extra-cellular pH and cerebral vascular resistances has been recently confirmed by WAHL (1970), inducing local charge of the pH of the fluid surrounding small arteries and observing modifications of the vessel diameter. The presence of a correlation between CBF and CSF pH, instead of between CBF and arterial CO_2 was observed in experiences of chronic induced respiratory modifications, such as chronic alkalosis and acidosis.

The results observed during respiratory alkalosis were contradictory and, at variance, responsible for criticism about the theory on the extracellular pH.

On the contrary, we were able to demonstrate a direct relationship between blood flow of the caudate nucleus of the brain of cat and CSF pH during chronic induced respiratory acidosis.

These results will be discussed in comparison with the results observed when using a different technique for measuring CBF.

These data were presented and discussed as a possible support of the theory on the pH of the extracellular fluid as the single factor responsible for the regulation of the caliber of cerebral vessels.

Reactivity of Smooth Muscles of Blood Vessels Under the Influence of Physical Factors

By D. Koradecka
The Central Institute for Labour Protection, Warszawa, Poland

Investigations of peripheral blood circulation tend to estimate the reactivity of particular circulation regions. Particularly important are here the responses to common thermal stimuli. Many papers have been devoted to these investigations, but there is still no clear discernment as far as the responses to cutaneous and muscular circulation stimuli are concerned.

Investigations were carried out in two groups of 30 persons each. The control group consisted of subjects not exposed professionally to noxious environmental factors. The experimental group was composed of workers professionally exposed to a noxious mechanical factor (vibration). Age means of the two groups were similar (group I - 29 years, group II - 32 years). Studies were performed in subjects lying supine in a room with ambient temperature 24 ± 2 ^{O}C, and atmospheric pressure of 752 ± 10 mmHg. The following investigation methods were applied: Occlusive and volumetric plethysmography by means of Elema-Schonander's plethysmograph, rheography with the aid of Schufried's rheograph, sphygmography by Boucke-Brecht's indicators, skin temperature registration using thermistor thermometry, capillaroscopy by Leitz-Wetzlar's capillaroscope, electrocardiography and blood pressure registration. Initial values were recorded after 20 minutes rest. Afterwards, forearms and hands were exposed to thermal stimuli. Water bath at 45 ^{O}C was applied as a hot stimulus acting till the skin temperature reached the value of + 34 ^{O}C. Thanks to this method initial states of subjects examined were unified before acting, during 10 minutes, by a cold stimulus being water bath at + 14 ^{O}C.

The action of thermal stimuli being over, the 20 minute period of restitution of circulation indexes was observed. In the picture are represented changes of blood circulation indexes in the forearm and hand during thermal stimulus action in the normal as well as exposed group. A, B, C, indexes give the blood circulation picture in the middle finger, while D index recorded simultaneously illustrates blood circulation conditions in the proximal segment of the forearm where muscular tissue constitutes 80 %. When applying heat locally the cutaneous blood flow and skin temperature of the fingers as well as the pulse volume increase. The pulse volume in the forearm, however, measured synchronously is reduced. This contrast in blood circulation responses in the forearm and fingers may be also observed when applying local cold, and in the course of restitution. Increase of the pulse volume in the forearm is concomitant with the decrease of cutaneous blood flow as well as of skin temperature. In the course of restitution the contrast in the reactivity of cutaneous circulation indexes in relation to muscular circulation is strongly marked, too. Identical investigations were carried out in a group exposed to a noxious mechanical factor being prolonged vibration in this case. Changes in indexes of cutaneous and muscular circulation which have been stated in this group during thermal stimulus action, however, appear to be similar on the contrary to the control group. Decrease of cutaneous blood circulation indexes is concomitant with the drop of muscular circulation ones. This fact provides evidence for the increase of vasoconstrictor tonus. A decrease of tonus of large muscular and elastic arteries in experimental group was stated simultaneously. Such a belief is supported by the following results of experiments: Increase of pulse volume, lowering of dicrotic notch, decrease of elastic resistance of circulation together with the extension of basal arterial oscilation.

These data allow to suppose that under the influence of prolonged exposure to a mechanical factor, disturbances of thermoregulating function of peripheral blood circulation occur as a result of disorders of the neurogenic regulation with a marked myogenic component in the vessels of the hand exposed directly to this factor.

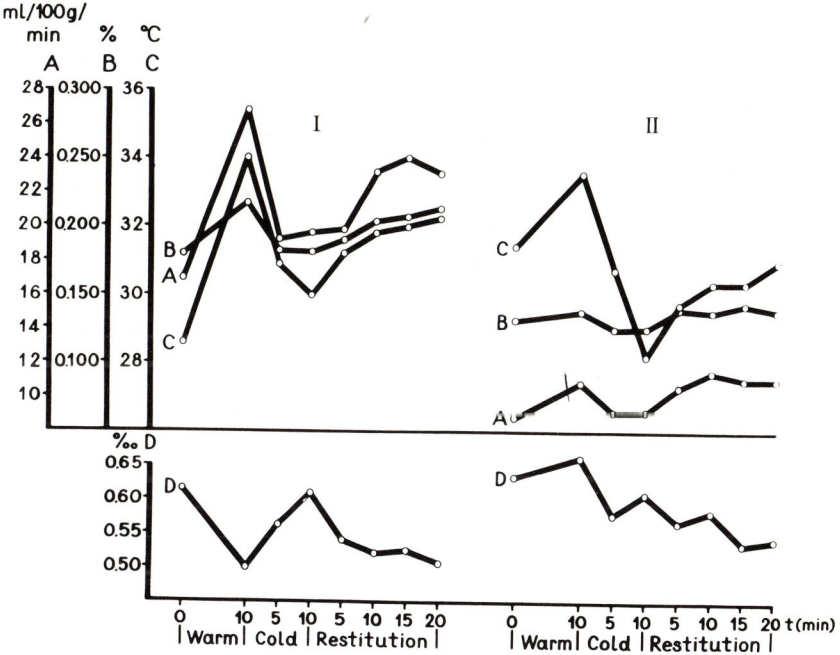

The peripheral circulation response to local thermal stimulation in normal subjects (I) and in those exposed to prolonged vibration (II).
Fingers: A - skin blood flow; B - pulse volume in per cent; C - skin temperature. Forearm: D - relative pulse volume by rheography.

Properties of Arterial Collateral Vessels

By J. Iriuchijima
Department of Physiology, Faculty of Medicine, University of Tokyo, Japan

In dogs anesthetized with pentobarbital, properties of the collateral vessels
for the common carotid and femoral arteries were studied by calculating collateral resistance from pressure and flow data.

1) During occlusion of either artery, as seen in the figure where the left common carotid flow (LCF), left intrasinus pressure (LSP) and systemic arterial
pressure (AP) were simultaneously recorded, left panel, collateral resistance
(r) gradually decreased in about 30 seconds, from initial collateral resistance
(ICR = 0.43 mmHg/ml/min in the figure) to steady state collateral resistance
(SCR = 0.24).
2) Intravenous injection of nitrites diminished ICR but not SCR, thus ICR
approached SCR. In the experiment reproduced in the figure, right panel,
ICR = SCR = 0.25 mmHg/ml/min, 15 minutes after administration of a long-
active nitrite, N-ethoxy-carbonyl-3-morpholinosydnonimine (SIN-10), at a
dose of 1 mg/kg. A similar but more evanescent effect was observed with
nitroglycerin (15 µg/kg). Nitrites dilated almost the same collateral vasculature which would gradually open during arterial occlusion.
3) During arterial occlusion, electrical stimulation of sympathetic vasocon-
strictors increased and that of sympathetic vasodilators decreased the pressure distal to the site of occlusion. This finding indicates that nerve stimulation affected peripheral resistance more markedly than collateral resistance
and further that the innervation of collateral vessels is rare, compared to
that of resistance vessels.

For the femoral constrictor response, the lumbar sympathetic chain was stimulated above the L 3 ganglion and, for the dilator response, the L 3 white
ramus was stimulated.

References

1) IRIUCHIJIMA, J., KOIKE, H.: Carotid flow, intrasinusal pressure, and
   collateral flow during carotid occlusion. Amer. J. Physiol. 218, 876-
   879 (1970).
2) IRIUCHIJIMA, J., KOIKE, H., KURIHARA, M.: Circuit parameters of
   carotid collaterals. Pflügers Arch. 322, 304-309 (1971).

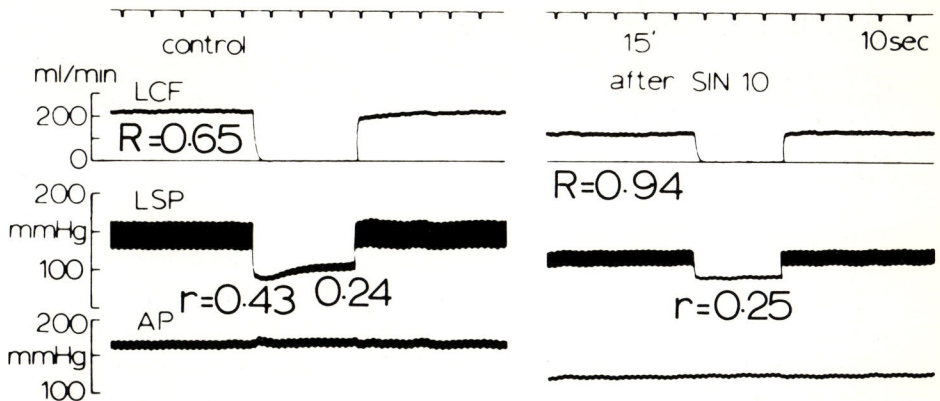

Summary and Discussion

Discussion of the paper of AGNOLI (BETZ, LASSEN, FIESCHI): The gradual adaptation of cerebral blood flow to prolonged exposure to respiratory acidosis or alkalosis gives an interesting clue to the regulation of cerebral blood vessel's tone. The evidence in favour of adaptation is still not entirely uniform, especially with regard to prolonged respiratory alkalosis, while the adaptation to prolonged hypercapnia in cats and rats seems well documented. This phenomenon brings in the first place the role of chemical enviromental factors, and is in line with the classical thinking that blood flow to the nervous tissue is essentially functional to and driven by its metabolic demands. The effective signal would be the H^+ ion concentration. However, it was mentioned that other ions might interact with the H^+ and by this interactions complicate the reactions in certain experimental or clinical condition. Due to stability of $[HCO_3^-]$ in the ECF of the brain, the main drive of the ECF pH are the respiratory variations, much as in the pCO_2 Severinghaus electrode. Just as in this electrode, if the $[HCO_3^-]$ in the medium is altered the sensor acts "as if" the pCO_2 varied in the opposite direction. This is essentially what happens in the course of adaptation, where a gradual increase of CSF $[HCO_3^-]$, the pCO_2 remaining high, is sensed as a gradual return to normal of $PaCO_2$: Indeed, the CSF pH and CBF vary in an almost parallel way. This observation fit well with the interpretation of some pathological findings (recently described with the ^{133}Xe clearance technique), such as the post-hypoxic reactive hyperemia (luxury perfusion), the vasomotor paralysis with loss of autoregulation and impaired or paradoxical responses to CO_2 (steal phenomena), the high blood flow in severe metabolic encephalopathies such as diabetic com.

In the discussion of the paper of Dr. IRIUCHIJIMA he was asked (LASSEN) about the influence of vasodilator drugs on the collateral vessels and he reported that he had produced dilatation with nitroglycerine. In a comment to Dr. IRIUCHIJIMA's paper Dr. GREEN pointed out that the method which was used did not differentiate between dilatation of the collateral vessels on the one hand and constriction of the peripheral vessels on the other.

Chairmen: C. Fieschi, I. Leusen

On the Fine Structure of the Portal Vein in Different Rodents

By F. Hammersen
Anatomisches Institut der Universität, Freiburg i. Br., Germany

As the portal vein of different rodents has become a widely used object for
experimental approaches to elucidate the functional behaviour of vascular
smooth muscle (1, 2, 3, 4, 5) some data pertaining to the correlative morphology of this specialized blood vessel should be presented:
1) All portal veins investigated so far (mouse, rat, guinea-pig, rabbit) are
characterized by an unusual thickness of their media and a typical stratifica-
tion of the total wall (fig. 1). Immediately subjacent to the endothelium lies
a thin layer of loosely packed and circularly arranged muscle cells which
constitute the "subendothelial" musculature. This is separated by a rather
broad band of connective tissue from the underlying media proper whose cells
are orientated longitudinally. The adjoining adventitia whose outer limits are
poorly defined contains vasa vasorum and bundless of unmyelinated axons,
but no muscle cells. When comparing the different thickness of the media in
these species it became apparent that its variations neither parallel the weight
of the animal nor the frequency of the spontaneous contractions, since the
smallest of our animals (mouse) shows the thinnest wall but the highest contractile activity, while the other three species exhibit a lesser activity with
a substantially thicker wall which is of rather the same size despite the considerable deviations in body weight and other parameters.
2) One of the characteristics of the media cells proper is a great number of
pleomorphic processes which are often free of myofilaments and whose membranes are studded with vesicles. These protrusions establish many close
membrane contacts of which some belong to the tight junction variety while
others seem to be gap junctions.
3) The distribution of the myofilaments seems to be less homogeneous and
their arrangement somewhat looser than in other vascular smooth muscle.
Dependant on the state and/or mode of contraction thick and thin filaments
could be distinguished with typical dense bodies in between which are attached
to the plasma membrane by means of simple hemidesmosomes. The high activity of this musculature is especially reflected by the large amount of mitochondria and free ribosomes and a well developed rough-surfaced endoplasmic reticulum.
4) Quite often round or oval electron lucent profiles of different size could be
found between the muscle cells either showing intimate relations to the latter
or lying completely isolated. These structures correspond to desintegrating
and detaching processes of smooth muscle cells which because of their appearance and their close relationship to the media cells could easily be misinter-
preted as axons making synaptical contact.
5) When comparing the distribution of the nerve and vasa vasorum in the portal vein one finds that they just reach the adventitio-media border in the mouse, while penetrating much further into the wall in the other species investigated (fig. 1). In the guinea-pig the nerves extend up to the subendothelial
connective tissue layer and are scattered throughout the longitudinal muscu-

lature in a similar fashion to that found in rabbits, while in the rat they are only located between, but not within, the two muscle layers. The axons of these small nerves exhibit typical richly vesiculated terminal beadings which display true neuro-muscular synapses predominantly in the guinea-pig, but gain a much closer contact with their effector cells in all the other three species than usually seen in vascular smooth muscle. In the nerve terminal areas the majority of the synaptical vesicles are of the small translucent variety combined with a few larger ones containing an electron dense core.

References

1) HOLMAN, M.E., KASBY, C.B., SUTHERS, M.B., WILSON, J.A.F.: J. Physiol. 196, 111 (1968).
2) VOTH, D., SCHIPP, R., AGSTEN, M., SCHÜRMAN, K., KOHLHARDT, M., DUDEK, J.: Arch. Kreisl.-Forsch. 60, 364 (1969).
3) GOLENHOFEN, K., LOH, v. D.: Pflügers Arch. 319, 82 (1970).
4) JOHANSSON, B., LJUNG, B., MALMFORS, T., OLSON, L.: Acta physiol. scand. 349, Suppl. 5 (1970).
5) LJUNG, B.: Acta physiol. scand. 349, Suppl. 33 (1970).

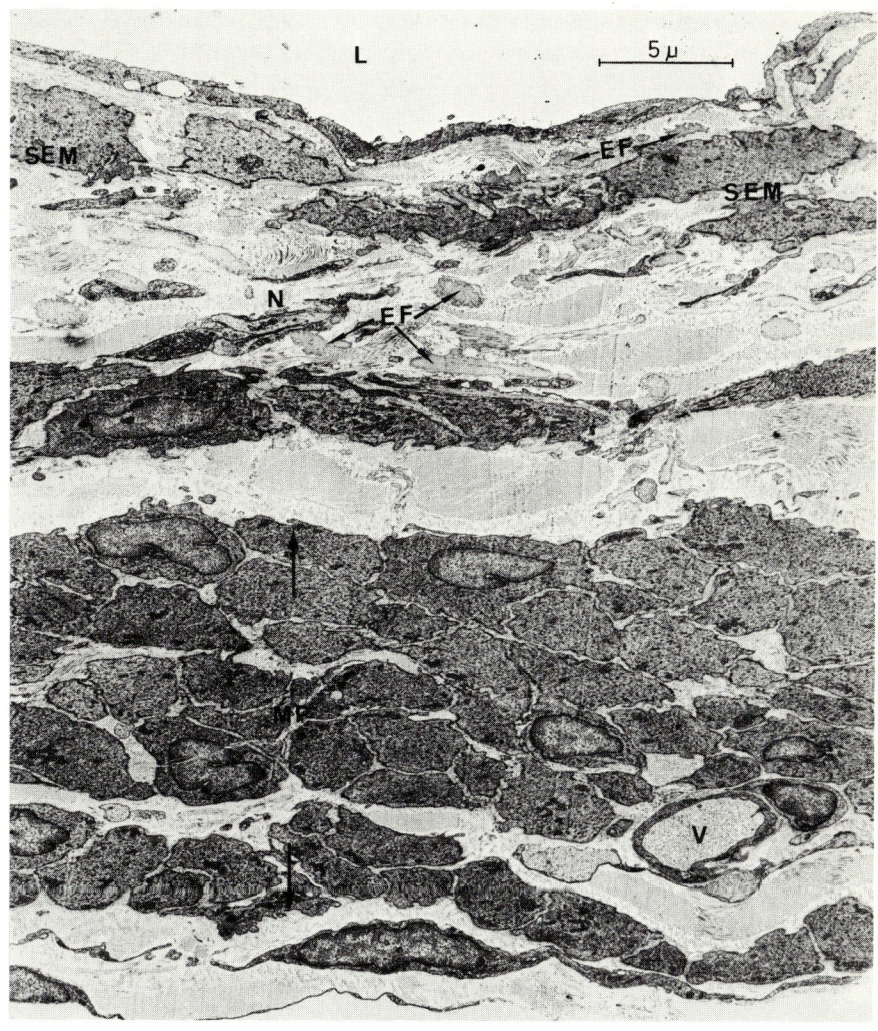

Cross-section of the rabbit's portal vein demonstrating the typical stratifi-
cation of the entire wall and the thick media proper (MP) with its muscle
cells orientated longitudinally.
L = Vascular lumen; EF = Elastic fibrils; N = Unmyelinated nerve; V = Vas
vasis; SEM = Subendothelial musculature. x 4800

Ultrastructural Cytochemistry of Coronary Smooth Muscle. An Electron Microscopic Observation in Human and Canine Coronary Arteries and Coronary Collaterals

By M. Borgers, J. Schaper, and W. Schaper
Cardiovascular Department, Janssen Pharmaceutica, Research Laboratoria, Beerse, Belgium

The topic of this study is to gain information about the localization of some non-specific and specific phosphatases known to be associated with important cell organelles of smooth muscle cells as plasma membranes, lysosomes, golgi apparatus, endoplasmic reticulum and nuclear envelopes and this in the coronary system of man and dog, and in the coronary collaterals of the dog after induction of growth.

Between normal human and canine coronary vessels only quantitative differences in enzyme equipment were noticed, the topographic distribution was essentially the same. To provoke the growth of the collateral system in the adult animal the following experimental model was used: An ameroid constrictor was placed around the left circumflex artery which occluded progressively the arterial lumen in a delay of 18 days.

As a result from this constriction a number of preexistant collateral junctions connecting the anterior descending artery with the circumflex artery, start growing. It appeared from previous studies (1) that during the transformation of these small arteriolar junctions important morphologic changes took place, and that these changes could be classified into well defined time-related stages
1) The early stage of alteration, situated 3 weeks after implantation of the constrictor, which is mainly characterized by signs of injury and repair present in every layer of the collateral wall.
2) The proliferative stages (8-12 weeks) during which a considerable thickening of the vascular wall is observed, together with the appearance of a subendothelial thickening.
3) The advanced stages (1-2 years) which are characterized by the reorganization of the media and of the intimal proliferative zone.

Some of the processes involved in the transformation, such as lysis, phagocytosis, connective tissue reaction, cell division, formation of new extracellular material, suggest active participation of hydrolitic enzymes.

The behaviour of the phosphatases in the collateral walls at different time-periods after implantation of the constrictor is summarized in table 1. These results are confirmed at the electron microscopic level. Some of these observations corroborate rather well the morphologic alterations present at the different stages of vessel wall transformation.

1) The morphologic changes appearing in the early periods of collateral growth are accompanied by a marked increase in ACPase activity. This enhancement is proportional to the degree of damage observed in the vessel wall.

2) Even under the hypoxic conditions presumed to be present in the early periods of development (a condition where one might expect a marked increase in 5' -nucleotidase activity) we failed to localize this activity in other cells than those who are situated in the adventitia.

3) The uniform distribution of activity towards ATP, ADP and TPP on the plasma membrane and in pinocytic vesicles of smooth muscle indicate that a polyphosphatase rather than specific phosphatases is responsable for splitting these substrates. The polyphosphatase concept of FREIMAN and KAPLAN is furthermore strengthened by the concomitant behaviour of activities towards ATP, ADP and TPP during vessel wall transformation.

4) Finally, the presumption that modified smooth muscle cells of the intimal thickening are functionning differently from normal smooth muscle cells of the media is supported by the dissimilarity in enzyme equipment. The loss of plasma membranebound polyphosphatase in the modified smooth muscle cells, concomitant to the appearance of prominent intracytoplasmic diphosphatase and glucose-6-phosphatase activities, suggest that these cells are actively involved in the synthesis and the secretion of elastin, collagen, basement membrane material and ground substance, rather than being involved in transport processes of cations and metabolites, necessary for relaxation-contraction mechanisms.

References

1) SCHAPER, W.: The Collateral Circulation of the Heart, Schaper, W. (Ed.). Amsterdam: North-Holland, 1971.

table 1: Enzyme activities[+] in small coronary arteries of control dogs and in coronary collaterals at various time intervals after the occlusion of one major artery. Light microscopic evaluation.
[+]activity is estimated visually and graded as follows: - = no activity; + = weak ++ = moderate; +++ = strong activity. endoth. = endothelial lining; int. thick. subendothelial thickening; med. smc = medial smooth muscular layer; adv. fibro. = adventitial fibroblasts and fibrocytes.

Enzyme	Localization	Control	3 weeks	8-12 weeks	1-2 years
Alkaline Pase	endoth.	-	-	-	-
	int. thick.			-	-
	med. smc	-	-	-	-
	adv. fibro.	-	-	-	-
Acid Pase	endoth.	+	+++	+	+
	int. thick.			++	+
	med. smc	+	+++	+	+
	adv. fibro.	+	+++	+	+
ATPase	endoth.	++	++	+++	++
	int. thick.			-	-
	med. smc	+++	-	+++	+++
	adv. fibro.	+++	+	+++	+++
ADPase	endoth.	++	++	++	++
	int. thick.			-	-
	med. smc	++	-	++	++
	adv. fibro.	+	+	+	+
TPPase	endoth.	+++	+++	+++	+++
	int. thick.			-	-
	med. smc	++	-	+++	++
	adv. fibro.	+	+	+	+
5'-nucleo-tidase	endoth.	-	-	-	-
	int. thick.			-	-
	med. smc	-	-	-	-
	adv. fibro.	++	+++	++	++
G-6-Pase	endoth.	+	+	++	++
	int. thick.			++	++
	med. smc	+	++	+	+
	adv. fibro.	++	+++	++	++

Sarcoplasmic Reticulum, Mitochondria and Filament Organization in Vascular Smooth Muscle

By A. P. Somlyo, C. E. Devine, and A. V. Somlyo
Department of Pathology, Presbyterian-University of Pennsylvania Medical Center, Philadelphia, Pennsylvania and the School of Medicine of the University of Pennsylvania, USA

The purpose of our studies was to characterize, in vascular smooth muscle (VSM), the sarcoplasmic reticulum (SR) and mitochondria that may contribute to the release and uptake of activator calcium during excitation-contraction coupling, and to determine the organization of myofilaments with particular attention to the state of myosin.

Studies with extracellular markers (ferritin, lanthanum) verified that the tubular structures identified as SR in the VSM of rabbit main pulmonary artery (MPA), portal-anterior mesenteric vein (MV) and small mesenteric arteries (300-500 /u diameter) were closed to the extracellular space. The SR formed fenestrations overlying surface vesicles. There were close (100 Å) contacts traversed by electron dense material between the SR and both non-specialized surface membrane and the surface vesicles. The volume of SR was greater in the MPA (although this estimate included some rough endoplasmic reticulum) than in the MV: And (at 25 °C) the MPA contracted when stimulated with drugs even after thirty minutes in 4 mM EGTA-Ca-free solution, while the rabbit MV did not. Maximal contractile responses of MPA in depolarizing, Ca-free (179 mM K, 4 mM EGTA) solution (at 25 °C) were greater to norepinephrine than to acetylcholine. Electron opaque accumulations of strontium were found in the SR of guinea pig and rabbit MV incubated, prior to fixation, in 10 mM Sr-Krebs solution, and in the SR of glycerinated rabbit MPA.

Mitochondria (up to 7 /u in length) of rabbit MV formed close relationships (average 44 Å) with the surface vesicles: When examined after relaxation from Barium-contractures (in Ca-free solution) massive electron opaque deposits of Ba were present in these mitochondria.

Transverse sections of rabbit MV showed quasi-rectangular arrays of thick (180 Å average diameter) filaments associated with regular arrays of thin (50-70 Å diameter) filaments in muscles fixed under physiological conditions. Thick filaments were also present in unstretched MV. A third type of filament (100-120 Å) associated with dense bodies and assumed to be analogous to the cytoskeleton of obliquely striated muscles, was also observed. In MV strips stretched beyond the contractile limit and incubated in hypertonic (with sucrose) solutions, large numbers of ribbon-like structures were observed in transverse sections: These appeared to be due to aggregation of the thick filaments.

Supported by NIH Grant HE 08226, NSF Grant GB 20478, the George L. and Emily McMichael Harrison Fund for Gynecological Research and the NIH General Research Support Grant FR 05610. A. P. S. is recipient of USPHS Research Career Program Award K3-17833.

We conclude that: 1) A sarcoplasmic reticulum that accumulates divalent cations and forms close couplings with the surface membranes is present in vascular smooth muscle, supporting the suggestion (1-6), that twitch contractions of spike generating smooth muscles are mediated by translocation of calcium from such SR;

2) The mitochondria of VSM can accumulate divalent cations capable of activating contraction, and the mitochondrial-surface vesicle contacts may serve as points of ion transfer;

3) The arrays of thick myofilaments (7-9) represent organized myosin and are consistent with a sliding filament mechanism of contraction in VSM, and

4) The pharmacomechanical coupling in Ca-free, depolarizing solution may be triggered at sites of SR contact with surface membranes, but is not readily explained by inward diffusion (2-3) of extracellular calcium. Release of Ca from centrally located SR (but continuous with regions forming couplings) could take place through propagation of an active response of Ca current along the SR membranes.

References

1) SOMLYO, A.V., SOMLYO, A.P.: J. Pharmacol. exp. Ther. 159, 129-145 (1968).
2) SOMLYO, A.P., SOMLYO, A.V.: Pharmacol. Rev. 20, 197-272 (1968).
3) SOMLYO, A.P., SOMLYO, A.V.: Pharmacol. Rev. 22, 249-353 (1970).
4) DEVINE, C.E., SOMLYO, A.P.: Fed. Proc. 29, 455 Abs. (1970).
5) SOMLYO, A.P., DEVINE, C.E., SOMLYO, A.V., NORTH, S.N.: J.Cell Biol. (in press).
6) SOMLYO, A.P., DEVINE, C.E., NORTH, S.N., SOMLYO, A.V.:Int. Cong. Physiol. Sci. Abstract, 1971.
7) DEVINE, C.E., SOMLYO, A.P.: J. Cell Biol. 49, 636-649 (1971).
8) RICE, R.V., McMANUS, G.M., DEVINE, C.E., SOMLYO, A.P.: Nature 231, 242-243 (1971).
9) SOMLYO, A.P., SOMLYO, A.V., DEVINE, C.E., RICE, R.V.: Nature 231, 243-246 (1971).

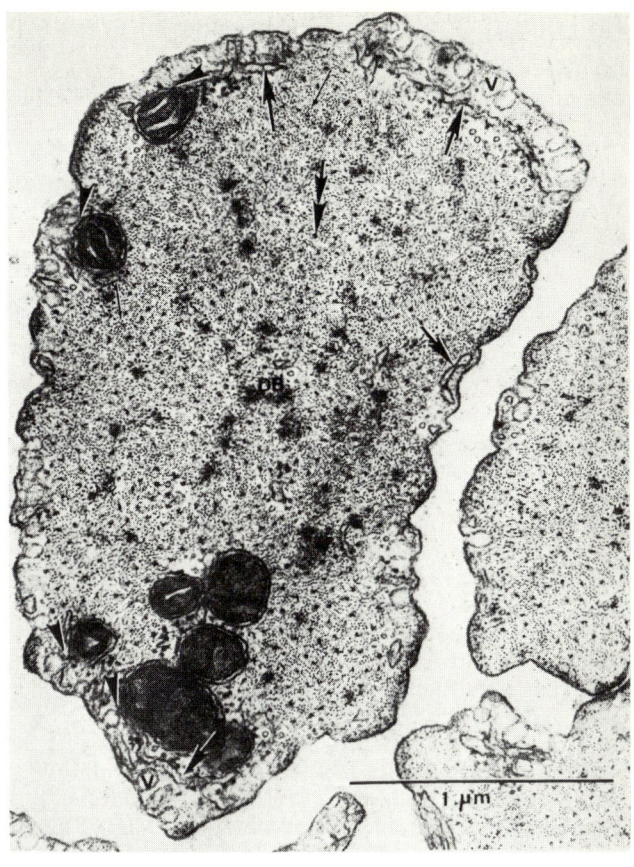

Transverse section through a smooth muscle cell of a rabbit portal-anterior mesenteric vein showing regular arrays of thick filaments (120-200 Å in diameter, double arrow) and thin filaments (50-80 Å in diameter, small arrow) with dense bodies (DB) scattered throughout the cytoplasm. Dense bodies attached to the cell membranes alternate with regions of surface vesicles (V) which are in close relationship with some of the mitochondria (arrow heads). Elements of the sarcoplasmic reticulum (large arrows) lie close to the surface vesicles or cell membrane. Tissue excised and stretched to its approximate in vivo length, incubated for 30 minutes in normal Krebs and fixed in cacodylate-buffered 2 % glutaraldehyde containing 4.5 % sucrose, post-fixed in 2 % osmium tetroxide in the same buffer, block-stained with uranyl acetate. Thin sections stained with alkaline lead citrate and viewed in a Hitachi HU 11 E electron microscope.

The Ultrastructure of Contractile Elements in Smooth Cells During Various Functional Stages

By D. Keyserlingk
II. Anatomisches Institut der Universität Berlin, Germany

The main problem encountered in the study of smooth musculature is the fixation of the cells in a relaxed condition. The smooth muscle cells are liable to contract when we manipulate the tissue mechanically, especially when we cut off the blood supply. A perfusion fixation with glutaraldehyde with and without formaldehyde therefore seems to be best method. Tissues with a minimum tendency to contract unspecifically are juvenile uterus and the uterus under the influence of progesterone. In these cases one finds almost only thin filaments over large cross-sectional areas. This applies for the autochthonous as well as for the musculature of the vessels. Thicker filaments are not found in longitudinal section either. After extraction with glycerine the contractile material consists mostly of thin 50-70 Å wide filaments which sometimes are closely packed together (fig. 1a). After contraction of the muscle cells of the vessels by incubating with ATP we find thick filaments appearing in a regular pattern and with a maximum diameter of 180-200 Å (fig. 1b). In these cases we found many thick filaments only in a contracted condition.

The situation is a little different in smooth muscle cells which contract unspecifically very easily e.g. intestinal musculature. In untreated intestinal muscle cells one finds very often thicker filaments, which are described by SHOENBERG (1964) as dark filaments. RICE et al. (1970) have also shown cross-sections of these cells, relaxed muscle cells - where no thicker filaments appeared. After extraction with glycerine one often finds dark filaments which are about twice as wide as the 50-70 Å wide thin filaments. The thick filaments which appear after incubation with ATP i.e. after contraction, distinguish themselves from these. They are in general wider, less dark and contain a fine inner structure. In super-contracted cells models, which will be dealt with later, one only finds the latter type of thick filaments. The special features of these filaments can be best seen in longitudinal section at greater magnifications. They are up to 0.6 μm long in our material. They have tapered ends in both directions. The thickest area in the middle is about 200-260 Å wide. In both directions the diameter falls to 100 Å. Statements concerning diameter are therefore of limited value in cross-section. Only these filaments correspond fully with the morphological criteria of NONOMURA (1968) and SHOENBERG (1969) in vitro for myosin filaments which are prepared from smooth muscle cells. I would once again like to demonstrate the difference between dark filaments and myosin filaments in an example which is just beginning to contract under the influence of ATP. We think it is very probable that the dark filaments change into the myosin filaments. Between the myosin filaments and the thin filaments one finds lateral bridges. A comment on relaxation with EDTA. Our glycerine extraction solution in which the tissue is placed immediately after removal contains EDTA. We also find single contracted muscle cells in the extracted material in spite of this. EDTA is therefore not capable of reversing a morphological change which has set in.

I would now like to deal with the small arteries of the intestinal wall in order to obtain a better comparison with the previous speech. After glycerin extraction we find dark filaments in the contractile material but no typical myosin filaments. Myosin filaments only appear after incubation with ATP and contraction. The course of contraction in glycerin-extracted cell models has already been described by us thoroughly in an earlier paper (1970). Actin and myosin filaments slide primarily parallel to each other. In the super-contracted stage the myosin filaments collect themselves in the middle of the cells. The actin filaments adhere for a while to the cell membrane, then they tear off and are pulled towards the myosin filaments. It is now very striking that the actin filaments are drawn towards the myosin filaments but they do not enter the accumulation of myosin filaments; they disappear. Therefore at the end of the contraction process no actin filaments are present in the super-contracted cells but only myosin filaments.

We come to the following assumption: During the contraction the myosin polymerises from lower aggregation forms to myosin filaments. The dark filaments can be seen as precursors and localisation of polymerisation. They probably consist of actin filaments wound round each other with attached myosin molecules. Simultaneously with the polymerisation of the myosin, the actin depolymerises. We can thus explain our results, which certainly demonstrate in part extreme situations, in this manner.

To end with I would just like to show that dark filaments and those which are similar to the myosin filaments appear in potassium contractures. When we fix by perfusion without pretreatment they are lacking in these vessel muscle cells.

References

1) KEYSERLINGK, D.: Z. Zellforsch. 111, 559 (1970).
2) NONOMURA, Y.: J. Cell Biol. 39, 741 (1968).
3) RICI, R.V., MOSES, J.A. et al.: J. Cell Biol. 47, 183 (1970).
4) SHOENBERG, C.F.: Int. Symp. Biochem. Gefäßwand Fribourg 1968. Angiologica 6, 233 (1969).

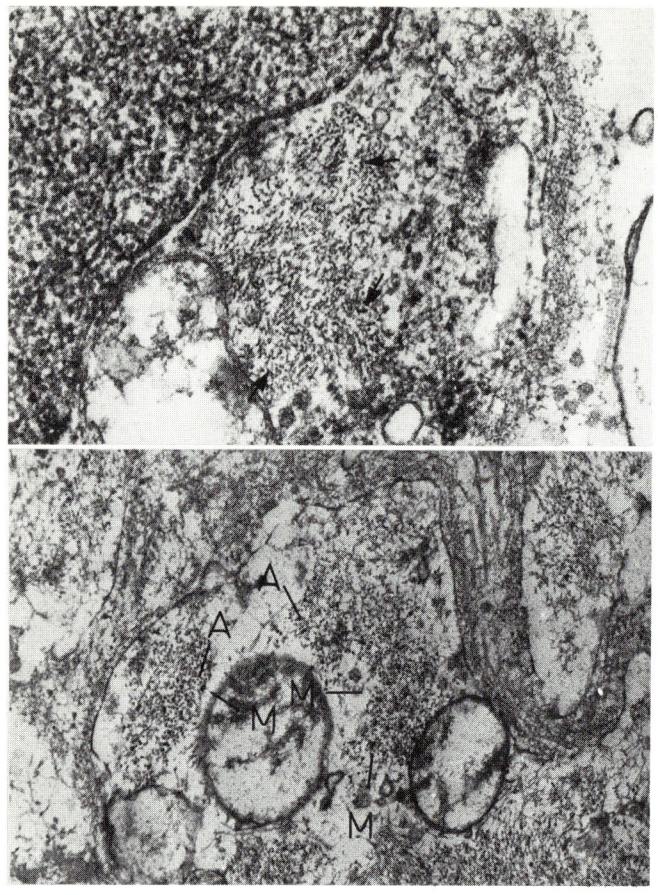

Cross-sections of the contractile elements of small arteries from the juvenil uterus. a) After glycerin-extraction only thin filaments, somtimes closely packed together (↓). b) After extraction and incubation with ATP, i.e. contraction, thick (M) and thin (A) filaments are found.
a) 90 000 : 1; b) 66 000 : 1

Summary and Discussion

Paper of HAMMERSEN: During the discussion of his presentation, Professor
HAMMERSEN emphasized that the distribution of nerves varies considerably
in different species and in different blood vessels, and hence, observations
on one species cannot be generalized, without experimental studies, to another species. In general, he found that the electron microscopic results are
in good agreement with fluorescense microscopic studies of innervation in
blood vessels. Regarding the question of myoneural synapses, he emphasized that, contrary to the notoriously great neuromuscular distances in many
other blood vessels, synaptic clefts of only 200 Å can be found in guina pig
portal vein: These seem to be the shortest neuromuscular distances reported
to date. In general, he found that the neuromuscular distance, in a given species, is shorter in the portal vein than in the other blood vessels examined
by him. Although he found that portal veins are generally well innervated, it
appeared that not every smooth muscle cell in the media receives its own
innervation, but only a certain fraction of the fibers. The structures through
which cell to cell propagation occurs in portal veins are presumably the ne-
xuses and/or gap junctions. The observations of both Professor HAMMER-
SEN's and Professor SOMLYO's laboratories concurred that the number of
mitochondria is often considerably greater in guinea pig than in rabbit portal
vein smooth muscle.

The discussion of Professor SOMLYO's presentation centered around the organization of myosin in vertebrate smooth muscle and the relative contribu-
tions of the surface vesicles, sarcoplasmic reticulum and mitochondria to
the control of calcium movement during contraction and relaxation. Professor
SOMLYO pointed out that the relatively regular organization of thick filaments
into a quasi-rectangular lattice with an approximately 700 Å filament to filament spacing and the observation that the subunits composing thick filaments
are smaller (25-50 Å) than the thin filament diameters are very much consistent with the view that the thick filaments represent the organized form of
myosin in mammalian smooth muscle, and are not aggregated thin filaments.
He has indicated that they have demonstrated thick filaments in both stretched
and in unstretched smooth muscles and that past difficulties in visualizing the
thick filaments were probably due to problems in preservation for electron
microscopy. He noted that extracellular markers (ferritin and lanthanum) enter the surface vesicles, and, therefore, these sites cannot be regarded as
intracellular calcium storage sites. In the reply to the question from Professor KEATINGE, he pointed out that the surface couplings of the sarcoplasmic reticulum with the cell membrane are observed with sufficient frequency
in the portal-anterior mesenteric vein to account for an electromechanical
coupling mechanism, at these sites, similar to the twitch mechanism at cardiac muscle couplings and skeletal muscle triads. He further added that the
ability of the sarcoplasmic reticulum to take up divalent cations that activate
contraction has been verified by the localization of strontium (a divalent cation that can functionally substitute for and is more electron opaque than calcium) in the sarcoplasmic reticulum (including the junctional SR forming the
couplings) of rabbit and guinea pig portal-anterior mesenteric veins and in
rabbit pulmonary arteries (SOMLYO, A.V. and SOMLYO, A.P., Science, in
press). During the discussion of mitochondrial participation in contractile regulation, he noted that some uncertainties remain, because while having suc-

ceeded in blocking the mitochondrial uptake of barium in rabbit portal-anterior mesenteric veins with oligomycin and cyanide, they did not simultaneously block relaxation in these preparations. He entertained, as possible mechanisms of relaxation under these conditions, a reduced transmembrane influx of barium or a reduced ATP supply for contraction due to the action of metabolic blockers, but emphasized that resolving this question will require further experiments.

Dr. KEYSERLINGK's interpretation of the organization of myosin in smooth muscle was at considerable variance from that of the previous speaker. In recognizing this controversy, the chairman pointed out that techniques of preservation for electron microscopy of smooth muscle having been sufficiently improved and a number of laboratories now pursuing these studies, it is probable that a common and generally accepted structure will be recognized within the next year.

Chairmen: A. P. Somlyo, F. Hammersen

Round Table Discussion. Moderator: Dr. O. Hudlicka

I. Introduction - Einleitung

KOEPCHEN:
Zu Beginn dieser gemeinsamen Tagung von Theoretikern, die sich mit der Klärung der Fundamentalprozesse an den Gefäßen befassen, und klinisch tätigen Angiologen erhebt sich die Frage nach Nutzen und praktischer Anwendbarkeit der Grundlagenforschung auf diesem Gebiet. Es soll darum hier kurz angedeutet werden, wo die Grundlagenforschung zur Zeit steht, welche Probleme in der Diskussion sind und wo die Lücken unserer Kenntnisse für ein Verständnis der Gefäßfunktion im Rahmen des Gesamtkreislaufes liegen.

Die Wirkungen der Gefäße auf den zentralen und peripheren Kreislauf ergeben sich aus ihrem wechselnden Kontraktionszustand. Der zugrundeliegende Fundamentalprozeß ist also der einer nervös gesteuerten Kontraktion der glatten Muskulatur der Gefäße. Die Kette der physiologischen Ereignisse ist dabei wie am Skelettmuskel die Ankunft von Nervenimpulsen, die Ausschüttung eines Überträgerstoffes an den Nervenendigungen, Beeinflussung der Permeabilität der Muskelmembran, Erregungsentstehung an der Muskelmembran und Kontraktion. Diese Vorgänge sind am Skelettmuskel besser bekannt und relativ einheitlich. Neben der Ähnlichkeit des Fundamentalprozesses existieren jedoch am glatten Muskel sehr wesentliche Abweichungen, die hier besonders interessieren. Die wesentliche Verschiedenheit bei der Kontraktion der Gefäßmuskulatur besteht a) in einer größeren Unabhängigkeit des Kontraktionsvorganges von der Innervation. So ist die Tätigkeit der Gefäßmuskulatur eine Interaktion zwischen einer, meist rhythmischen Eigentätigkeit der glatten Muskulatur und der Gefäßinnervation; b)die Reaktionsformen der Gefäßmuskulatur sind weit variabler als die der Skelettmuskulatur und sehr wesentlich vom jeweils untersuchten Objekt abhängig.

Die wesentlichen Themen der Grundlagenforschung auf diesem Gebiet sind zur Zeit die Fragen der Erregungsentstehung in der glatten Muskulatur, ihrer Fortleitung und der Koppelung zwischen elektrischen und mechanischen Phänomenen. Ein wesentlicher Fortschritt in der Erforschung der Membranprozesse ergab sich durch die Anwendung der Technik der intrazellulären Ableitung auch an der glatten Gefäßmuskulatur.

Die Ergebnisse und Probleme nach dem hier diskutierten Stand der Grundlagenforschung lassen sich, soweit dies in der Kürze überhaupt möglich ist, etwa folgendermaßen umreißen: Zentralnervöse Ursache des Gefäßtonus ist ein dauernder Impulsstrom in sympathischen Nerven, die Gesetzmäßigkeiten der Entstehung des sympathischen Tonus auf supraspinaler und spinaler Ebene werden zur Zeit intensiv erforscht, auf Einzelheiten kann hier nicht eingegangen werden.

Als Folge dieser sympathischen Erregungen wird an den Nervenendigungen Noradrenalin freigesetzt. Dieses bewirkt eine Depolarisation der Gefäßmuskelzellen. Es erfolgt eine Änderung der Permeabilität, gefolgt von einem Ioneneinstrom, der wiederum durch Bereitstellung intrazellulärer freier Kalziumionen den Kontraktions- und damit Konstriktionsvorgang auslöst. Daneben ist das Noradrenalin offensichtlich in der Lage, an bestimmten Gefäßen

Kontraktionen auch ohne elektrische Membranveränderungen einzuleiten. Wie weit solche direkten Wirkungen des Noradrenalins auf die intrazelluläre Kalziumbereitstellung einen physiologisch bedeutenden Mechanismus darstellen, wird diskutiert.

Eine große Reihe von Untersuchungen liegt über die Spontanaktivität isolierter Gefäße vor, was vor allem durch die Methodik der Untersuchung isolierter Gefäßabschnitte bedingt ist. Der Ausgangspunkt dieser nicht nerval ausgelösten Spontanaktivität liegt in "Schrittmacherzellen", die an wechselnden Stellen eines Gefäßes auftreten können. Form und Ausmaß der Kontraktion hängt von Frequenz und Synchronisationsart dieser Schrittmacher ab. Hochfrequente Schrittmacheraktivität hat tetanusartige tonische Kontraktionen zur Folge. Die Periodendauern solcher Spontanrhythmen bewegen sich in verschiedenen Bereichen von Sekunden bis zu Stunden. Oft sind mehrere solcher Rhythmen einander überlagert. Elektrische Korrelate im Membranpotential sind langsame Potentialschwankungen oder, oft diesen aufgesetzt, schnelle Spike-artige Potentialverläufe. Die Spike-ähnlichen Potentiale lösen Kontraktionen aus, solche wurden jedoch auch ohne Spikes korreliert zu langsameren Potentialschwankungen beobachtet. Die elektrischen Phänomene sind in verschiedenen Gefäßabschnitten verschieden.

Die in diesem Zusammenhang noch offenen Probleme sind folgende: Gibt es rein tonische langsame Gefäßkontraktionen oder sind diese immer durch hochfrequente Erregungsbildung nach Art eines Tetanus hervorgerufen?

Gibt es spezielle Schrittmacherzellen, von denen die myogenen Erregungen ausgehen oder ist jede Gefäßmuskelzelle potentiell zur Erregungsbildung in der Lage?

Was kann an der Gefäßmuskulatur überhaupt als "Aktionspotential" bezeichnet werden, unterscheiden sich die langsamen und die schnellen Potentialabläufe in ihrem ionalen Mechanismus voneinander?

Wann ist Na^+, wann Ca^{++}, wann sind beide Ionen Träger der elektrischen Potentialschwankungen?

Neben spontanrhythmischen, mechanischen und elektrischen Phänomenen werden rhythmische Schwankungen der Ionentransporte von Na^+ und K^+ beobachtet. Hier ist die Frage, ob es sich dabei um die Ursache oder um Begleiterscheinungen der Spontanrhythmik handelt und ob die Schwankungen der Ionentransporte in rhythmisch schwankender Energiebereitstellung durch den Zellstoffwechsel oder in Permeabilitätsschwankungen der Membran begründet sind.

Insgesamt läßt sich also feststellen, daß auch die Fundamentalprozesse der Kontraktionsentstehung an der Gefäßmuskulatur noch bei weitem nicht geklärt sind. Entscheidende Lücken unserer Kenntnisse, die zum Zweck einer Anwendung auf den Kreislauf ausgefüllt werden müßten, sind vor allem folgende:

1. Die Fundamentaleigenschaften wurden fast ausschließlich am isolierten, denervierten Gefäß untersucht. Wir wissen noch wenig über das Zusammenspiel mit der natürlichen Gefäßinnervation.

2. Bedingt durch die methodischen Schwierigkeiten, vor allem der intra zellulären Ableitung, wurden bisher fast ausschließlich die großen Gefäße untersucht, bevorzugte Präparate sind Aorta, Arteria carotis und Vena portae. Dagegen spielen sich die für die Regulation des peripheren Kreislaufes entscheidenen Prozesse ja an den kleinen Gefäßen ab. Ob die an den großen Gefäßen gewonnenen Erkenntnisse auf die peripheren Gefäße übertragen werden können, ist äußerst fraglich, da schon zwischen den bisher verwendeten Präparaten erhebliche Unterschiede bestehen und ebensolche Unterschiede zwischen zentralen und peripheren Gefäßabschnitten zu erwarten sind.

Um an das in der Eröffnungsansprache unseres Herrn Ehrenvorsitzenden gebrauchte Bild von den Raumschiffen anzuknüpfen: Die Verbindung der klinischen Erfahrungen mit der Grundlagenforschung ist zwar das Ziel sowohl der Kliniker wie der Theoretiker, aber bis zur erfolgreichen "Ankoppelung" ist noch ein weiter Weg zu überwinden.

II. Factors Initiating Vascular Smooth Muscle Contraction

GOLENHOFEN:
Electrical activity in vascular smooth muscle is not a single phenomenon: there are true action potentials, similar to spikes known from the skeletal muscle, there are vessels with pacemaker potentials as vena portae, and there are still slower electrical changes (second or minute rhythms occuring at intervals of seconds or minutes) which could trigger the faster ones. The terminal vessels are much more similar to the portal vein, with its spontaneous activity, where electrical activation of mechanical activity is essential, than to large arteries where non-electrical activation of contraction is more important.

KEATINGE:
In some blood vessels, at least, both electrical and non-electrical activation of contraction is of great importance and electrical activation has a different function and significance in different blood vessels. Some vessels, for instance, the portal vein and turtle aorta, have spontaneous electrical and mechanical activity similar to the heart and the main function of these blood vessels is to propel blood. In other vessels electrical activation plays no part. The sheep carotid artery gives smooth non-rhythmical activity and the important means of activation of the contraction is non-electrical activation by noradrenaline and by calcium, which is allowed to enter the cells or which can be released from calcium stores; but even in these arteries, electrical activation plays a very important part in response of the vessels to noradrenaline released from the nerve fibres. Action potentials which would occur in this case would spread to a limited distance through the vessel wall; but when the vessel is exposed to a sustained high concentration of noradrenaline only brief bursts of electrical activity occur and the vessel remains contracted with a cessation of electrical activity.

BOHR:
Although electrical activation appears in a number of vessels, any vascular smooth muscle can contract without it: it depolarizes when placed in potassium chloride solution, contracts in response to angiotensin and norepinephrine and contracts in high concentrations of calcium without action potentials.

Agents causing relaxation affect even depolarized vessels where action potentials cannot be responsible for relaxation. So there are non-electrical means for activation of contraction in the vascular smooth muscle.

WEZLER:
Vascular smooth muscle does not behave in any different way than any other smooth muscle. It has local depolarization and spikes, it has propagated depolarization, and the differences between vena portae and aorta or carotid artery are only gradual and methodological ones, depending on the number of registrating electrodes and the chance with which the electrode is placed into a cell which is active.

SOMLYO:
Differences between vascular smooth muscles which generate action potentials and those which do not are not only methodological ones. The sucrose gap technique applied to the portal vein and pulmonary artery in the same species - rabbit - has shown spontaneously generated action potentials in the portal vein with a frequency which increased in response to excitatory drugs, while the pulmonary artery responded with a graded depoliarzation. It was possible to record action potentials even in the pulmonary artery in one out of 35 fibres, but these action potentials were completely different than those in the portal vein and were not associated with contraction.

Both electromechanical and pharmacomechanical coupling plays a role in vascular smooth muscle contraction; the role is different in various smooth muscles and depends on the degree of excitation.

III. The Role of Transmitters and Sympathetic Innervation in the Control of Cerebral Blood Flow. Specific Sensitivity of Cerebral Vessels to pCO_2

LASSEN:
Cerebral blood vessels respond generally very little, if at all, to catecholamines administered intravascularily, in spite of some controversial findings presented in this meeting. Whenever the cerebral circulation was completely isolated from the extracerebral cephalic circulation, there was no response to catecholamines. This is, teleologically speaking, very important, for instance, in shock when large amounts of catecholamines are present in blood, because the lack of reactivity to catecholamines in cerebral vessels preserves normal circulation in the brain. It would appear that catecholamines cannot penetrate the capillary endothelium, and autoradiography failed indeed to demonstrate any labelled noradrenaline in the wall of cerebral blood vessels when it was administered intravascularily. Small constrictions could be observed in pial vessels - which are representative for the whole cerebral vascular bed - when catecholamines were applied locally, but the concentration used was 1000 or even 10000 times higher than those found under physiological conditions. So again the lack of reactivity to noradrenaline could be predominantly a question of permeability.

ROACH:
The lack of reactivity of cerebral blood vessels to catecholamines does not seem to be a question of permeability. When noradrenaline was applied directly into the muscle cells in the basilar artery or middle cerebral artery, it

elicited hardly any contraction, while both the carotid artery and mesenteric artery contracted very vigorously. So the question arises whether the muscle cells in cerebral vessels are different, or whether their innervation is different.

JOHANSSON:
In the portal vein only one muscle layer is innervated and alpha receptors are present only in this innervated muscle layer. The other layers of smooth muscle cells are excited in this vessel by the spreading of myogenic activity. There are no nerves in intracerebral blood vessels, and no nerve endings which would stimulate the growth of receptor substance.

KEATINGE:
On the other hand, the inner muscle layer in the sheep carotid artery, which is not innervated, is 10-100 times more sensitive to noradrenaline. So the presence of nerves is not a prerequisitive for the development of receptor sites and sensitivity to catecholamines.

SOMLYO:
Innervation is not an absolute requirement for alpha adrenergic responsiveness to transmitters. Vascular smooth muscles in the umbilical vessels, which are not innervated at all, respond to alpha adrenergic stimuli, and, on the other hand, larger coronary arteries and larger arteries to skeletal muscle, or small pulmonary arteries in birds, are rather insensitive to catecholamines similarily as cerebral blood vessels.

BOHR:
The unresponsiveness to catecholamines is not dependent on innervation. Helical strips of cerebral arteries do not respond to catecholamines, though they do respond to depolarizing doses of potassium chloride.

FLECKENSTEIN:
There is no need for any special effect of lack of effect of catecholamines since vascular tone is dependent on calcium concentration in dependence on pH (see KNABE in this meeting). The increase in pH causes an increase in the vascular tone because it activates calcium; the decrease in pH causes vasodilatation because calcium loses its vasoconstrictor effect in low pH and it cannot activate ATPase. Once the pH is over neutral, very small amounts of calcium are sufficient for activation of ATPase and ATP splitting.

HUDLICKA:
Since activation of ATPase and splitting of ATP are basically the same in all vessels, there are evidently some specific qualities in cerebral blood vessels which make them unresponsive to catecholamines and very sensitive to changes in pCO_2.

FIESCHI:
The great sensitivity of cerebral blood vessels to the changes in pCO_2 dilatation in response to hypercapnia and constriction in response to hypocapnia - is indeed very specific. Similar responses could be found, for instance, in hind limb blood vessels only if they were denervated; but cerebral blood vessels can adapt to a very high pCO_2 - their diameter can return to normal af-

ter they have been exposed to a high pCO_2 even if the pCO_2 remains high. It is clear today that it is actually the extravascular pH (which can be changed very rapidly according to the intravascular changes of pCO_2) which is decisive for the reaction of the cerebral blood vessels. The concentration of the extravascular bicarbonates is rather constant under normal circumstances, and only when it changes, as in posthypoxic acidosis, the vessels respond. In an acute stage of brain disease, the concentration of bicarbonate may change and this could explain why, under such conditions, brain vessels may respond by vasodilatation to vasoconstrictor drugs.

IV. The Role of Cyclic AMP in the Process of Contraction of Vascular Smooth Muscle

LASZT:
ATP plays a crucial role in the process of contraction in the muscle cell being split into ADP and inorganic phosphate while energy is being released for contraction. ATP can also be split immediately into AMP and two molecules of phosphoric acid by adenylcyclase which, at the same time, connects the phosphoric acid to ribose, forming cyclic AMP. The content of cyclic AMP in the vascular smooth muscle is then regulated by adenylcyclase on one side and by phosphodiesterase, which splits cyclic AMP hydrolytically into AMP, on the other side. The reactivity of vessels is dependent on the content of these two enzymes. Towards periphery the content of adenosincyclase increases and the content of phosphodiesterase decreases and, thus, the content of cyclic AMP increases. Catecholamines increase the activity of adenylcyclase and this increase could be depressed by beta-blocking agents. It has been suggested that adenylcyclase may act as a beta-receptor and that the vessels may become dilated by setting free the cyclic AMP which could act as the second messenger. Two findings are, however, contradictory to this suggestion. Firstly, the ability of vessels to contract increases towards periphery. Secondly, cyclic AMP causes a very small vasodilatation in coronary vessels and only if applied in very large doses. This lack of response is probably not due to a low permeability of blood vessels to cyclic AMP because labelled cyclic AMP penetrates the vessels.

SOMLYO:
Following criteria for cyclic AMP to act as a second messenger in the vascular smooth muscle have been fulfilled.

1. The phosphodiesterase inhibitor - theophyllin - potentiates the effect of the beta-adrenergic agent, isoproterenol, on the membrane of the vascular smooth muscle in a concentration which has no effect on the membrane potential by itself, therefore its effect is not an additive one but a true potentiation

2. Dibutyryl cyclic AMP, which penetrates the cells better than cyclic AMP itself, has been shown to mimic the action of isoproterenol on the vascular smooth muscle membrane potential, including its dependence on the external potassium concentration.

3. The concentration of cyclic nucleotides rises in vascular smooth muscles stimulated by beta-adrenergic stimulating agents.

This preliminary evidence seems to be reasonably convincing in that cyclic AMP plays a role as an inhibitory second messenger in vascular smooth muscle.

V. The Phenomenon of Microvascular Hypersensitivity

BALOURDAS:
I would like to draw the attention of the panelists and all the participants in this Symposium to the phenomenon of the hypersensitivity of the vascular smooth muscles and the hyperreactivity of the respective microvessels. We have found this phenomenon of microvascular hypersensitivity-hyperreactivity in our experimental investigations on various categories of experimentation pertaining to the changes of microcirculation induced by pharmacological agents and by pathophysiological states and conditions experimentally produced.

In all these in vivo experiments by the mesocaecum bioassay and the direct visualization of the rat mesoappendiceal microvessels we observed structural and functional changes, i.e. microvascular hypersensitivity-hyperreactivity of the arteriolar-precapillary sphincters and micro-angiopathy. The hypersensitivity was found by the epinephrine threshold vasoconstrictor test. The hypersensitive microvessels become hyperreactive to various endogenous and exogenous vasoactive agents and substances circulating within the microvascular beds.

The vascular hypersensitivity-hyperreactivity (hereditary, acquired) may explain many morbid entities and conditions in biology and medicine: In angiospastic diseases, Raynaud's syndrome, Buerger's disease, periarteritis nodosa, necrotizing arteritis, toxemia in pregnancy with hypertension and preeclampsia and eclampsia, in essential hypertension, and other conditions. In essential hypertension (spontaneously hypertensive rats) we found remarkable hypersensitivity-hyperreactivity of the mesoappendiceal and retinal microvessels without microangiopathy.

Therefore, the microvascular hypersensitivity might be the pathogenetic aetiology of essential hypertension initiated by many hereditary and acquired factors. Many morbid entities in biomedicine may be precipitated and aggravated by the hyperreactivity of the microvessels and the microangiopathy with all the clinical implications and complications.

Die venöse Druck-Volumencharakteristik in der Bewegung bei Gesunden und bei pathologischen Gefäßveränderungen

Von H. Rieckert und H. Loeprecht
Abteilung für Physiologie und Department für Chirurgie, Abteilung II der
Universität Ulm, Germany

Die Druck-Volumencharakteristik des venösen Gefäßsystems der unteren Ex
tremität zeigt das elastische Verhalten und die Stabilität der Gefäßwand bei
Erhöhung des Transmuraldruckes. Methodisch wurde bisher diese Druck-
Volumenrelation mit Hilfe der Druck-Volumenschleife (RIECKERT und PAU-
SCHINGER, 1969) registriert. Eine Staumanschette wird am Oberschenkel
kontinuierlich auf systolischen Druck aufgepumpt und in gleicher Geschwindigkeit wieder abgelassen. Der Manschettendruck, der praktisch dem Venen
druck identisch ist und das zugehörige Weichteilvolumen, das segmentple-
thysmographisch registriert wird, werden gleichzeitig auf die Koordinaten
eines x/y-Schreibers aufgezeichnet. Es ergibt sich eine schleifenförmige
Kurve, die einer Druck-Volumenregistrierung bei wechselndem Übergang
vom Liegen zum Stehen entspricht. Diese Methode simuliert praktisch eine
orthostatische Belastung. Sie ist aber ein rein statisches Verfahren.

Wir haben deshalb versucht, die Druck-Volumencharakteristik auch als dyna
mische Methode in der Bewegung beim Gehen und Laufen aufzuzeichnen. Den
in der Bewegung werden pathologische Gefäßalterationen manifest. Der Venendruck wird über ein Stathamelement Sb 37 in einer Fußvene auf die y-Ach
das Wadenvolumen plethysmographisch auf die x-Achse eines x/y-Schreibers
registriert. Die Schleifenform wird durch die Muskelbewegung modifiziert.
Druck und Volumen nehmen im Gehen ab (Abb. 1).

Wie verändert sich diese Kurve z. B. bei Varikosis? Die Kurvenform ist bre
ter und nach oben gezogen. Es wird ein Volumen abgepresst, das aber in Re
lation zum Ausgangsvolumen gering ist. Die Kurve verschiebt sich nach klei
nerem Volumen, jedoch findet sich kein Druckabfall.

In den letzten Jahren sind die Beckenvenenthrombosen vor allem wegen der
jetzt günstigeren Operationsaussichten mehr in den Vordergrund getreten.
Auch hier bieten sich über die Druck-Volumenregistrierung in der Bewegung
oder im Drainageversuch nach einer artifiziellen Stauung Diagnosemöglich-
keiten. Abb. 2 zeigt die Schleifen bei einer Patientin mit Beckenvenenthrombose und Klappeninsuffizienz mit primärer Druck- und Volumenzunahme. De
Druck fällt sekundär wahrscheinlich über eine dehnungsbedingte Widerstands
abnahme im venösen Gefäßsystem gering ab. Die Drainagezeit nach Stauung
ist verzögert.

Postoperativ nach saphena-bypass können kurzfristige arteriovenöse Anastomosen angelegt werden um eine Thrombose im Transplantat zu verhindern.
Bei Klappeninsuffizienz kann der hohe Venendruck nach distal durchschlagen
und beim Liegen Druckwerte erreichen, die sonst nur im Fußbereich im Stehen vorkommen. Dadurch erhöht sich nach dem Starling'schen Gesetz die
Oedembereitschaft, welche nach Verschluß der Fistel völlig zurückgeht.

Methodisch sollten immer beide Größen, Druck und Volumen, registriert werden. In der Volumenregistrierung zeigen sich plethysmographisch abgenommen Artefakte, die aber bei Kenntnis der Methode eliminiert werden können. Günstig scheint hier die induktive Registrierung nach LITTMANN (1970). Der Druckaufnehmer sollte klein und gegen Lageänderung unempfindlich sein.

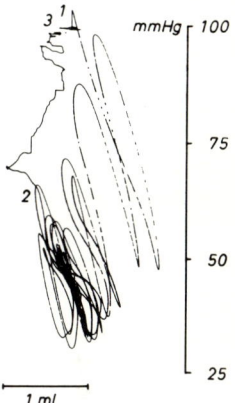

Abb. 1. Druck-Volumenschleife beim Gehen

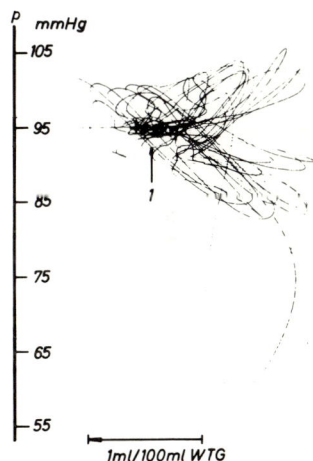

Abb. 2. Druck-Volumenschleife bei Beckenvenenthrombose

Untersuchungen zum gestörten Mesenchymstoffwechsel bei Thrombophlebitis

Von W. H. Hauss, G. Junge-Hülsing, H. Wagner und H. Möbitz
Medizinische Klinik und Poliklinik der Universität Münster/Westf., Germany

Unsere klinischen und experimentellen Untersuchungen zur Arteriosklerose-entstehung haben gezeigt, daß als initialer Prozeß eine Störung des Mesenchymstoffwechsels in der Arterienwand nachweisbar ist. Der Mesenchymstoffwechsel der Gefäßwand reagiert auf vielerlei Noxen regelmäßig mit einer Acceleration, die wir als unspezifische Mesenchymreaktion bezeichnen (1).

In einer weiteren Untersuchungsreihe sollte der Mesenchymstoffwechsel gesunder (Gruppe I) und thrombophlebitischer (Gruppe II) Venen (Vena femoralis dextra und sinistra) verschiedener Altersgruppen überprüft werden. Die gesunden Venen wurden von gefäßgesunden Verstorbenen entnommen; makroskopisch konnte an diesen Venen kein pathologischer Befund erhoben werden. In der Gruppe II waren bei den thrombophlebitischen Venen regelmäßig Thromben der Venenwand nachweisbar. Bei dieser Gruppe wurden die Gefäßsegmente in einer Entfernung von 5 cm distal bzw. proximal des Thrombus verwendet. Als Parameter zur Kontrolle des Mesenchymstoffwechsels diente der ^{35}S-Sulfat-Einbau in vitro in die sulfatierten Mucopolysaccharide der Venenwand nach von uns früher beschriebener Methodik (2).

In Übereinstimmung mit unseren früheren Untersuchungen zu Altersveränderungen im Mesenchymstoffwechsel von Aorten ließ sich auch im Mesenchymstoffwechsel der Venenwand ein typischer Altersgang nachweisen. Mit steigendem Lebensalter nimmt die Intensität des Sulfomucopolysaccharidstoffwechsels in der Venenwand ab (Abb. 1).

Thrombophlebitische Venen zeigten im Vergleich zu gesunden Venen zu allen untersuchten Lebensaltern einen um das Mehrfache gesteigerten ^{35}S-Sulfat-einbau in die sulfatierten Mucopolysaccharide der Venenwand. Dieser Untersuchungsbefund konnte mit einem t = 8,058 bei einem p < 0,001 statistisch gesichert werden (Abb. 1).

Die Ergebnisse unserer Untersuchungen weisen daraufhin, daß für die Entstehung einer Venenthrombose neben Veränderungen der Gerinnungsfähigkeit des Blutes sowie der Zahl und der Agglutinationsfähigkeit der Blutplättchen insbesondere eine Störung des Gefäßwandstoffwechsels von primärer Bedeutung ist.

Mit dankenswerter Unterstützung durch das Landesamt für Forschung NRW.

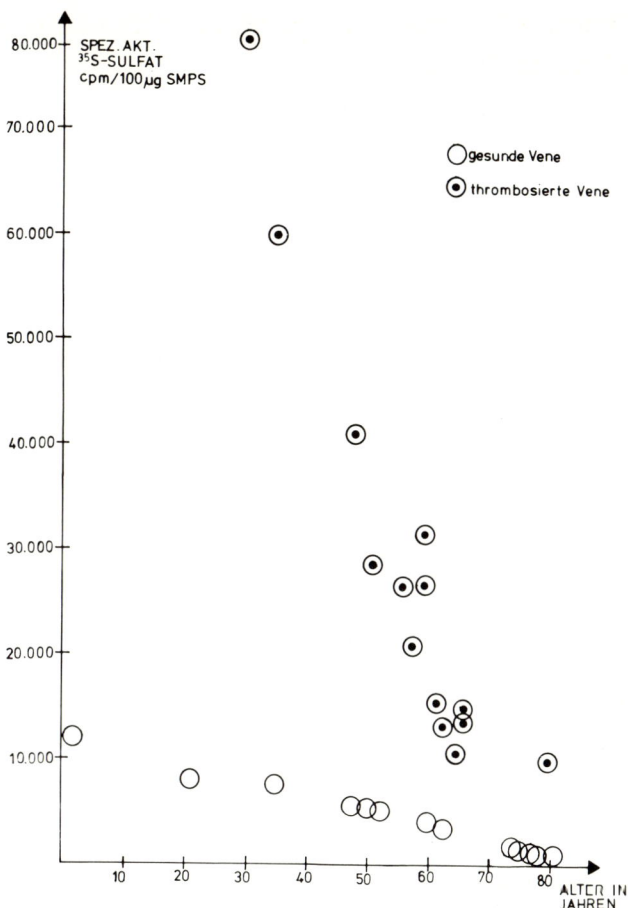

^{35}S-Sulfateinbau in die sulfatierten Mucopolysaccharide der Venenwand

Literatur

1) HAUSS, W.H., JUNGE-HÜLSING, G., GERLACH, U.: Die unspezifische Mesenchymreaktion. Stuttgart: Thieme, 1968.
2) JUNGE-HÜLSING, G.: Untersuchungen zur Pathophysiologie des Bindegewebes. Heidelberg: A. Hüthig, 1965.

Effect of Hypoxia on Reactivity of Cutaneous and Mesenteric Veins of the Dog

By P. Vanhoutte and I. Leusen
Laboratory of Normal and Pathological Physiology, State University of Ghent,
Belgium

The reactivity of spiral strips of saphenous veins and of longitudinal strips of mesenteric veins of the dog was investigated in an organ bath. Changes in isometric tension of the preparations and variations in PO_2 of the Krebs-Ringer solution were recorded. Contractions of the preparations were obtained by transmural electric stimulation (VANHOUTTE et al., 1967) or by addition of catecholamines to the bath solution. Anoxia ($PO_2 \leqslant 1$ mmHg) depressed the spontaneous activity exhibited by the mesenteric veins. Prolonged (30 to 60 minutes) hypoxia ($PO_2 = 40$ mmHg) slightly depressed the frequency-response characteristics to electric stimulation in the mesenteric but not in the saphenous preparations, while prolonged anoxia ($PO_2 \leqslant 1$ mmHg) depressed the reactivity of both types of veins, especially as regards the lower stimulation frequencies (1 to 2 cps). A short anoxia period (5 minutes) imposed during sustained contractions, evoked by either electric stimulation (1 to 10 cps adrenaline ($5 \cdot 10^{-7}$ g/ml) or noradrenaline ($5 \cdot 10^{-7}$ to 10^{-6} g/ml), depressed the reactions by approximatively 30 % (fig. 1). Those inhibitory effects of anoxia were greatly facilitated in glucose-free solutions. Prolonged anoxia in glucose-free solutions abolished the reactions to electric stimulation in both saphenous and mesenteric veins. The data indicate that, in isolated vein strips, the lack of oxygen decreases the myogenic activity, and to a lesser extent the reactivity to adrenergic stimulation in a way similar to what has been described for other vascular smooth muscle (FURCHGOTT, 1955; 1966; KEATINGE, 1964; LLOYD, 1967; SHIBATA and BRIGGS, 1967; DETAR and BOHR, 1968). Our experiments further suggest that, as for the arterial preparations (KEATINGE, 1964; FURCHGOTT, 1955; 1966; LUNDHOLM and MOHME-LUNDHOLM, 1965; SHIBATA and BRIGGS, 1967; NEEDLEMAN and BLEHM, 1970; SCOTT et al., 1970), anaerobic glycolysis plays an important role in the energy supply for the contraction of venous smooth muscle.

References

1) DETAR, R., BOHR, D. F.: Amer. J. Physiol. 214, 241-244 (1968).
2) FURCHGOTT, R. F.: Pharmacol. Rev. 7, 183-265 (1955).
3) FURCHGOTT, R. F.: Bull. N.Y. Acad. Med. 42, 996-1006 (1966).
4) KEATINGE, W. R.: J. Physiol. 174, 184-205 (1964).
5) KIRK, J. E., EFFERSE, P. G., CHIANG, S. P.: J. Geront. 9, 10-35 (1954
6) LLOYD, T. C., jr.: J. appl. Physiol. 22, 1101-1109 (1967).
7) LUNDHOLM, L., MOHME-LUNDHOLM, E.: Acta physiol. scand. 64, 275-282 (1965).
8) NEEDLEMAN, P., BLEHM, D. J.: Life Sci. 9, 1181-1189 (1970).
9) SCOTT, R. F., MORRISON, E. S., KROMS, M.: Amer. J. Physiol. 219, 1363-1365 (1970).
10) SHIBATA, S., BRIGGS, A. H.: Amer. J. Physiol. 212, 981-984 (1967).

11) SOMLYO, A.P., SOMLYO, A.V.: Pharmacol. Rev. **22**, 249-353 (1970).
12) VANHOUTTE, P., CLEMENT, D., LEUSEN, I.: Arch. int. Physiol.
Biochem. **75**, 641-657 (1967).

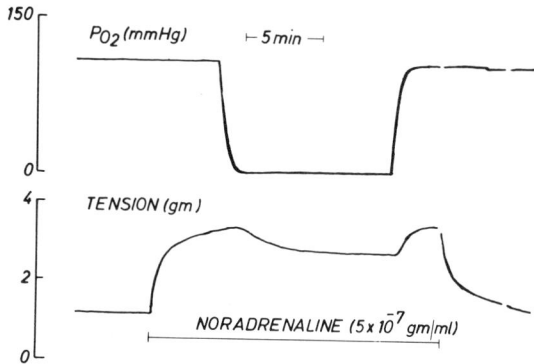

Effect of anoxia on a sustained contraction of an isolated saphenous strip.
Original record

Comparison of Veno-Vasomotoric Reactions in the Abdominal Circulatory Systems of Intestine, Stomach, Spleen, Liver and of the Kidney

By J. Lutz and J. Biester
Physiologisches Institut der Universität Würzburg, Germany

The veno-vasomotoric reaction (VVR) is the constrictive response of the resistance vessels to elevations of venous pressure. A constriction transfered
on a nervous-reflectory way can be excluded today with certainty; thus, the
response on the distension can be regarded with greatest probability as being
induced myogenly, that means the smooth muscle acts as a receptor and as
an effector, though not necessarily simultaneously in all sections. We have
examined the VVR in different abdominal circulatory systems of the cat by
perfusion with the animals own blood; here the constant flow perfusion had

proved to be best.

The most intensive reactions could be elicited in the liver, but only in the hepatic artery, where they were evoked by pressure elevations in the inferior caval vein or in the portal vein. So the proportion R_2/R_0 (resistance during the reaction divided by the control resistance) during a pressure increase of 18.4 mmHg in the caval vein amounted to 1.94 ± 0.10. The vessels of the portal vein on the other hand always stayed passive, the quotient R_2/R_0 being constantly below 1. If we plot down the reactive arterial pressure increase (ΔP_A) in relation to the eliciting venous pressure increase (ΔP_V) in a diagram, we get a figure, which contains all the active responses above the straight line $\Delta P_A = \Delta P_V$. Beyond the venous pressure elevations of 13.3 mmHg the intensity of the reactions does not increase in the same degree; A certain saturation appears. There are different conditions in the intestinal circulation, where up to 33 mmHg the venous pressure elevations are met with a continually increasing reply.

In the circulatory systems of spleen and stomach the appearance of the VVR could also be demonstrated in a qualitative and quantitative aspect (2,3). For the most important vascular beds, drained together into the portal vein, a comparison of the responses could be made with a combination of three independent perfusion systems (occlusive roller pumps). A venous pressure elevation of 24.2 mmHg in the mean resulted in a R_2/R_0 value of 1.25 ± 0.03 for the stomach, 1.42 ± 0.03 for the spleen and 1.49 ± 0.06 for the intestine. Between the reactions of spleen and intestine no significant difference exists, though it does between both of them and those of the stomach. The weaker response of the stomach might be attributed to the fact, that a part of the left gastro-epiploic artery was not fully perfused in contrast to the other gastric vessels, because of the contemporary perfusion of the spleen and the strong isolation of the vessels. Under perfusion with dextran only the values of the spleen stayed > 1; with application of papaverine (final concentration 0.2 mg/ml) all values approached one another. That the VVR stayed positive for a longer time in the spleen under dextran could be due to the fact, that it takes longer to wash out this organ from rests of plasma. A constrictive factor, however, according to our opinion is at all times necessary to enable baryno-genic reactions.

For the vessels of the kidney it was important to prove venously elicited constrictions, since various authors have described only dilatations after venous pressure increases (1,4,5). These were attributed partly to the large increment of tissue pressure and thus to the diminished transmural pressure. The increase of blood flow was also regarded as an effect of ureteral pressure elevation and a consequently reduced absorption of Na^+. So mechanisms exist in the kidney, which mask or diminish the VVR. We have got, however, unambiguous results of vasoconstrictions during the duration of venous pressure elevations of 22-50 mmHg in the sense of a VVR. Mechanical effects could be excluded by suppression under application of papaverine. On the other hand an intensification appeared in weak responses with small infusions of noradrenaline (already with local concentrations of $6 \cdot 10^{-8}$ g/ml), an intensification which we have also seen in the intestinal vascular bed. This too illustrates the significance of a basal tone for the responses of the vessels.

In the diagram values of the kidney are plotted down, also those under papaverine. Although the number of tests with kidneys still is small, the obviously weaker degree of response contrasts with the other results. The low reaction of renal vessels could suggest, that the eliciting pressure changes reach the place of action - possibly the afferent artery - only diminished. Besides the other mentioned mechanisms are to be considered. The described experiments indicate the different intensity of the VVR in certain circulatory systems under perfusing conditions and show that a locally different sensitivity - also towards disturbance - seems to exist in this general response of vessels.

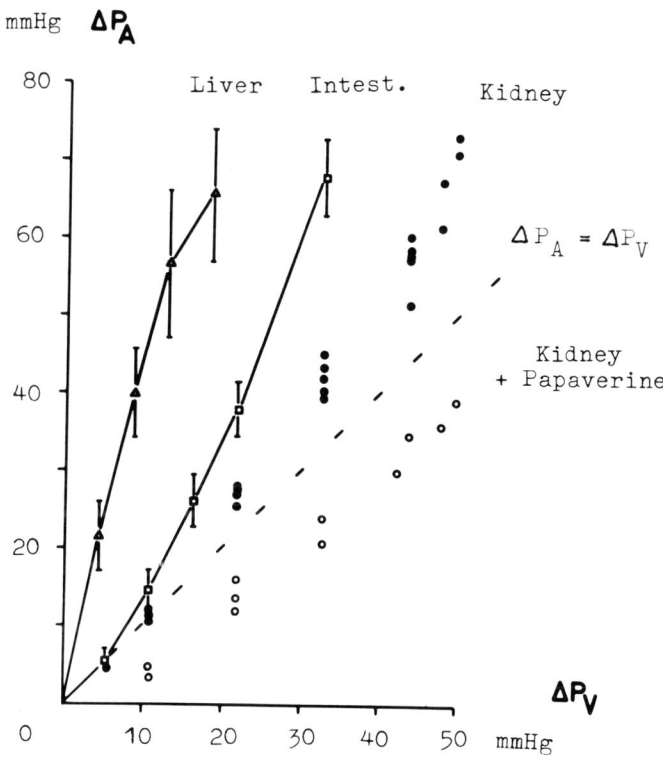

Comparison of the different degree of veno-vasomotoric reactions in some vascular beds

References

1) HINSHAW, L. B., BRAKE, C. M., IAMPETRO, P. F., EMERSON, T. E.: Amer. J. Physiol. 204, 119 (1963).
2) LUTZ, J., BIESTER, J.: Proc. 25th Int. Congr. Physiol. Scienc. Munich 19
3) LUTZ, J., HENRICH, H., PEIPER, U., BAUEREISEN, E.: Pflügers Arch. 313, 271 (1969).
4) TAKEUCHI, J., KUBO, T., SAWADA, T., FUNAKI, E., SANADA, M., KITAGAWA, T., NAKADA, Y.: Jap. Heart J. 6, 243 (1965).
5) THURAU, K., HENNE, G.: Pflügers Arch. ges. Physiol. 279, 156 (1964)

The Effect of Progesterone and Oestradiol-17B on the Spontaneous Activity of Isolated Venous Smooth Muscle

By P. B. Johnston and I. C. Roddie
Department of Physiology, The Queen's University of Belfast, N. Ireland

Longitudinal strips of bovine mesenteric vein were set up in organ baths and when perfused with a modified Ringer's solution showed vigorous irregular spontaneous activity, each contraction lasting about 2 minutes. Experimental strips were exposed to oestradiol-17B and progesterone dissolved in ethylene glycol in doses ranging from 0.5 to 10 µg/ml and control strips were simultaneously exposed to the vehicle alone. Both hormones depressed the mean tone and the amplitude of the contractions of the experimental strips. Though the changes with the 0.5 µg/ml doses were small, 10 µg/ml of eithe hormone caused a 50 % reduction in resting venous tone and almost complete suppression of spontaneous activity. It is concluded that oestradiol-17B and progesterone are venodilator substances in vitro.

Regional Peripheral Resistance in Experimental Hypertension

By H. Dahners, W. Breull, D. Kikis, D. Redel, J. Schotte, K. Stoepel and H. Flohr
Physiologisches Institut der Universität Bonn, Germany

Experimental arterial hypertension is generally linked to an overall augmentation of total peripheral resistance. It is, however, unknown how this increase in resistance is distributed over the various vascular beds of the systemic circulation and whether or not the different regions contribute to the same extent to the overall increase in vascular resistance. The distribution pattern of regional peripheral resistance was therefore studied in three forms of experimental hypertension in rats using the particle distribution method.

Methods

Three forms of experimental hypertension were investigated:
1) In the first group of 32 animals the right renal artery was constricted by a silver clip 4-6 weeks prior to the measurement. Systolic blood pressure was about 188 mmHg.
2) In the second group of 47 animals the left kidney was removed and the right renal artery was constricted as in the first group. Systolic blood pressure was about 216 mmHg.
3) In the fourth group of 31 animals hypertension was produced by DOCA-implantation and unilateral nephrectomy, and replacing the drinking water by 1 % saline. Average systolic blood pressure was 198 mmHg.
4) 66 untreated normotensive animals of the same strain served as control group.

During a slight ether anaesthesia a calibrated amount of J-131 labelled macroaggregated albumin (10 μC) was injected by direct puncture of the left ventricle. 4-5 minutes after the indicator injection the animals were sacrificed. 29 organs and tissues were removed immediately, weighed and homogenized. The radioactivity found in each organ was determined in a 3"-well type scintillation counter (FHT 660 B1).

The following values were calculated:
1) The fraction of cardiac output passing through each organ or tissue from the fraction of indicator found in each region.
2) Relative regional flow by dividing this fraction by the relative organ weight.

The data obtained for the different groups were analyzed by means of the χ^2-test.

Results

The results are summarized in table 1. The distribution pattern of cardiac output in hypertensive rats differs significantly from that found in normotensive rats and is different in the different forms of hypertension investigated.

Conclusions

It can be concluded from our data that renal and DOCA-hypertension are characterized by a significant change in the distribution pattern of regional peripheral resistance, local resistances contributing to a different extent to the increase of total peripheral resistance.

table 1. 100 X Fraction of CO reaching some organs in normotensive and three types of hypertensive rats.

	Normotensive		Goldblatt I			Goldblatt II			DOCA		
	Mean	SD	Mean	SD	P	Mean	SD	P	Mean	SD	P
Heart	4.9	2.0	7.1	2.9	.000 25	6.3	2.5	.002 5	5.9	2.6	.05
Lung	4.5	3.4	5.1	2.8	.000 25	5.2	3.9	.002 5	5.8	4.1	.025
Brain	4.8	2.2	5.0	1.9		4.8	1.6		3.9	1.5	.012 5
Liver art.	5.4	3.1	5.1	3.5	.012 5	5.4	2.3		3.6	1.9	.000 5
Liver tot.	21.5	6.4	26.1	7.4	.000 25	19.8	6.9	.002 5	29.3	10.4	.000 25
Kidney (s)	9.3	3.1	5.5	2.0	.000 25	5.9	2.4	.025	8.1	3.1	.000 25

Diskussion

FISCHER: Die Kurven von RIECKERT und LOEPRECHT zeigen, daß im venösen System nicht der Druck im Vordergrund steht, sondern daß er eher eine Funktion des Volumens darstellt: Nur wenn die Muskelpumpe (unter einem Kompressionsverband) wiederhergestellt wird, sinkt auch der Druck beim Gehen wieder ab. Ödem entsteht hierbei also nicht durch Drucksteigerung, sondern durch Stauung.

Kinetische Energie wird im venösen System erst durch die Muskelpumpe erzeugt und kann sich demnach auch erst jenseits derselben auswirken, falls kein Strömungshindernis besteht, welches zur Rückstauung (mit Blutanschoppung) und dann auch zur Steigerung des Druckes führt. Die vorgestellten künstlichen a-v-Fisteln zeigen eine weitere Möglichkeit der artifiziellen Drucksteigerung im Venensystem mit konkurrierendem Abstrom wie bei Schwangerschaftsvarikose: "Hochwasser im Hauptstrom staut den Nebenfluß und umgekehrt".

Die Beziehung Volumen-Druck zeigt sich besonders deutlich nach Arbeit: Erst wenn das venöse Volumen von der arteriellen Seite her wieder aufgefüllt ist, steigt auch der Druck an.

SCHNEIDER: Entscheidend für die Druck-Volumenbeziehung im Venensystem ist letztendlich die Funktion der Speicherung und Strömung, die nur bei niederen Drucken vollzogen werden kann - im Gegensatz zum arteriellen Hochdrucksystem, welches das Blut leitet und verteilt - und insofern können weder Druck noch Volumen isoliert verstanden werden, sondern nur in ihrem jeweiligen und jeweils besonderen Zusammenwirken.

Nach den vorgetragenen Befunden von HAUSS und Mitarb. wäre anzunehmen, daß die verstärkte Einlagerung von Mukopolysacchariden in der Gefäßwand durch vermehrte Neubildung "am Ort" zustande käme. Bei der Vasculitis dagegen, bei der wir eine ähnliche, wenn nicht sogar stärkere Imbibierung der Intima mit polychromatischen Substanzen nachgewiesen haben, scheinen diese aus dem Plasma zu stammen, die Endothelschranke ist dabei erheblich geschädigt. Da aber auch bei der Thrombophlebitis eine Endotheldesquamation festzustellen ist, erhebt sich die Frage, ob die MPS-Einlagerung bei den beiden genannten Krankheitszuständen auf vergleichbaren bzw. analogen Vorgängen beruht.

FISCHER fragt, ob die Steigerung des Mesenchymstoffwechsels nicht durch die phlebitische Entzündungsreaktion bedingt ist.

SUNDER-PLASSMANN weist darauf hin, daß der Begriff der allgemeinen Mesenchymreaktion von ihm zuerst verwandt wurde (Zbl. Chir. 1948, S. 326), und daß die Gefäßwand als mesenchymales Organ auch bei der Thrombophlebitis nur mesenchymal reagieren könne.

Chairman: W. Schneider

The Physiological and Pathophysiological Reactivity of the Microvascular Smooth Muscles and the Implications thereof in Biomedicine

By T. A. Balourdas
Howard University School of Medicine, Washington, D. C., USA

The microvascular beds through the smooth muscles containing α - and β-receptors respond to endogenous and exogenous vasoactive agents and exhibit excitement and relaxation of the respective vessels. The arteriolar-precapillary sphincters influenced by the various vasoactive agonists and antagonists and substances cause vasoconstriction and vasodilatation for the maintenance of the normotensive homeostasis and tissue functionings. The rhythmic contraction and relaxation (vasomotion) is regulated by (1) the tonic activity of vasomotor center via nervous impulses transmitted by the Autonomic Nervous System, (2) hormonal substances and stimuli, and (3) autoregulation by virtue of local metabolic activities. Using the mesocaecum bioassay preparation for direct visualization of the rat mesoappendiceal microvessels via biomicroscopy, experimental studies were undertaken pertinent to the changes in microcirculation induced by pharmacological agents and by pathophysiological states and conditions. Results: In all these in vivo experimental studies functional and structural microvascular lesions were observed, viz. microangiopathy and microvascular hypersensitivity-hyperreactivity. The hypersensitivity was tested by the Epinephrine vasoconstrictor threshold concentration Test of the arteriolar-precapillary sphincters and consisted of the remarkable reduction of the threshold from the average normal $1:4 \cdot 10^{-6}$ to $1:0.2 \cdot 10^{-9}$ and $1:0.1 \cdot 10^{-9}$ E.T.C. The development of microvascular hypersensitivity was observed in many experimental studies induced by drugs and by pathophysiological states. The hypersensitive microvessels become hyperreactive to normally or abnormally discharged and circulating endogenous catecholamines and other vasoactive agents. The vascular hyperreactivity (hereditary, acquired) may explain many morbid conditions: Angiospastic states, Raynaud's syndrome, Buerger's disease, periarteritis nodosa, necrotizing arteritis, essential hypertension, eclampsia in pregnancy, etc. Therefore, the Microvascular Hypersensitivity and Microangiopathy Theory, inferred from our experimental studies, may explain many morbid entities and clinical implications in Biomedicine.

References

1) BALOURDAS, T.A.: Proc. IV Int. Congr. of Pharmacology, Basel (Switzerland), July 1969, p. 382.
2) BALOURDAS, T.A.: Fed. Proc. 26, 682 (1967).

Elektrolytstoffwechsel von Erythrozyten, Aorta und Herz- und Skelettmuskulatur bei Ratten mit erblicher spontaner Hypertonie und DOCA-Hochdruck

Von F. Wessels und H. Losse
Medizinische Universitäts-Poliklinik, Münster/Westf., Germany

Klinische und experimentelle Beobachtungen in den letzten Jahren lassen keinen Zweifel an der Bedeutung des Elektrolythaushaltes in der Pathogenese einiger Hochdruckformen, insbesondere der essentiellen Hypertonie (2). Auch die Befunde unserer Arbeitsgruppe sprechen dafür, daß bei der essentiellen Hypertonie eine angeborene weitgehend generalisierte Störung des Natriumhaushaltes eine entscheidende Rolle spielt (6, 7, 8).

Um weiteren Einblick in die Zusammenhänge zwischen Elektrolythaushalt und Hochdruckentwicklung zu erhalten, führten wir Elektrolytstoffwechselanalysen verschiedener Gewebe an Ratten durch, die entweder an einer erblichen spontanen Hypertonie oder an einem DOCA-Hochdruck litten.

Untersuchungsgut und Methodik: Zur Erzeugung eines DOCA-Hochdrucks verabreichten wir 6-8 Wochen alten männlichen Wistar-II-Ratten eine 1 % NaCl-haltige Trinklösung sowie am 4. und 10. Behandlungstag je 50 mg/kg K. -Gew. eines DOCA-Depot-Präparates (Depot-Cortiron). Eine erbliche spontane Hypertonie mit systolischen Blutdruckwerten zwischen 160 und 190 mmHg erhielten wir durch selektive Inzucht über mehrere Generationen. 16 Tieren dieses Rattenstammes verabreichten wir 10 Tage lang Furosemid in einer Tagesdosis von 5 mg/kg K. -Gew. subcutan. Der Blutdruck wurde plethysmographisch am wachen Tier gemessen. Zur Analyse der Natriumkonzentration sowie des passiven Natriumtransportes der Erythrozyten benutzten wir eine bereits früher beschriebene Methode (5, 6, 8). Die Natrium- und Kaliumkonzentration sowie den Wassergehalt der Aorta, der Herz- und Skelettmuskulatur bestimmten wir nach der von ZUMKLEY (9) angegebenen Methodik, maßen dabei aber außerdem die Größe des Extrazellulärraumes in den verschiedenen Geweben mit ^{51}Cr-EDTA.

Ergebnisse: Bei Ratten mit spontaner Hypertonie sind passiver Natriumtransport und Natriumkonzentration der Erythrozyten gegenüber normotonen Kontrolltieren nicht verändert. Dagegen findet sich nach Kochsalzbelastung und Gabe von DOCA eine deutliche positive Korrelation zwischen Blutdruckhöhe und Natriumeinstrom in die Erythrozyten (r = 0, 324 , P< 0, 001, n = 125). Die Ergebnisse unserer Untersuchungen des Elektrolyt- und Wassergehaltes sowie der Größe des Intra- und Extrazellulärraumes von Aorta, Herz- und Skelettmuskulatur bei Ratten mit spontaner Hypertonie bzw. DOCA-Hochdruck lassen sich der Tab. 1 entnehmen.

Besprechung: Nach den sehr ausführlichen Untersuchungen der Arbeitsgruppe um FRIEDMAN (Übersicht bei 3) haben sowohl der intra-extrazelluläre Kalium- als auch der extra-intrazelluläre Natriumquotient einen erheblichen Einfluß auf Tonus und Reagibilität der Gefäßmuskulatur. Diese Quotienten aber sind bei beiden von uns untersuchten experimentellen Hochdruckformen in der

Aorta erniedrigt, ein Befund, der sowohl den Hypertonus als auch die bekann te Steigerung der vaskulären Empfindlichkeit gegen Pressorsubstanzen erklären könnte. Der gleiche Wirkungsmechanismus dürfte am ehesten auch Ursache der Blutdrucksenkung bei mit Furosemid behandelten Hochdrucktieren sein, bei denen unsere Untersuchungen in der Aorta eine deutliche Abnahme der intrazellulären Natriumkonzentration ergaben.

Andererseits muß einschränkend bemerkt werden, daß glatte Muskelfasern auf jede Dehnung mit einer Zunahme der Spontanaktivität reagieren (4), die ihrerseits zu einem Anstieg der Natrium- und Abnahme der Kaliumkonzentra tion im Intrazellulärraum führt (1). Damit könnte aber auch die Blutdrucker-höhung selbst für die von uns beobachteten Elektrolytverschiebungen in der Aorta verantwortlich sein. Bei derartigen hochdruckbedingten Elektrolytver-schiebungen sollte man jedoch erwarten, daß diese weitgehend auf die Gefäße beschränkt bleiben. Wir fanden dagegen bei beiden Hochdruckformen als Hin-weis auf eine weitgehend generalisierte Elektrolytstoffwechselstörung auch in Herz- und Skelettmuskulatur mehr oder weniger deutliche Kationen- oder Wasserverschiebungen sowie beim DOCA-Hochdruck eine Beschleunigung des passiven Natriumtransportes der Erythrozyten. Damit sind diese Befunde ei-ne weitere Bestätigung unserer Auffassung, nach der in der Pathogenese der essentiellen Hypertonie des Menschen die von uns nachgewiesene generali-sierte Natriumstoffwechselstörung eine wesentliche Rolle spielt.

Literatur

1) DAWKINS, O., BOHR, D.E.: Amer. J. Physiol. 199, 28 (1960).
2) LOSSE, H., WESSELS, F.: Med. Klin. 65, 815 (1970).
3) MERTZ, D.P.: Elektrolytstoffwechsel und arterielle Hypertonie. Stuttgart New York: F.K. Schattauer 1971.
4) SCHATZMANN, H.J.: Ergebn. Physiol. 55 (1964).
5) WESSELS, F.: Z. klin. Chem. 8, 278 (1970).
6) WESSELS, F., JUNGE-HÜLSING, G., LOSSE, H.: Z. Kreisl.-Forsch. 56, 374 (1967).
7) WESSELS, F., ZUMKLEY, H.: Z. Kreisl.-Forsch. 59, 427 (1970).
8) WESSELS, F., ZUMKLEY, H., LOSSE, H.: Z. Kreisl.-Forsch. 59, 415 (1970).

	Kontrolltiere	spontane Hypertonie		
		unbehandelt	unt. Furosemid	DOCA-Hochdr.
Gruppe	I	II	III	IV
Aorta				
H_2O (Gew. %)	$68,56 \pm 0,51$ (25)	$67,00 \pm 0,37$ (33)* ++	$67,52 \pm 0,83$ (15)	$66,42 \pm 0,67$ (16)* +
E Z (l/kg Fr.-Gew.)	$0,5505 \pm 0,0082$ (25)	$0,5050 \pm 0,0079$ (33)* +++	$0,5133 \pm 0,0076$ (15)* ++	$0,5106 \pm 0,0084$ (16)* ++
I Z (l/kg Fr.-Gew.)	$0,147 \pm 0,0083$ (25)	$0,178 \pm 0,0089$ (33)* +++	$0,164 \pm 0,0108$ (15)	$0,161 \pm 0,0077$ (16)* +
$[Na_i^+]$ (mval/IZ)	$29,93 \pm 11,36$ (21)	$39,65 \pm 5,96$ (28)	$10,95 \pm 9,35$ (15)** +++	$43,57 \pm 8,34$ (15)
$[K_i^+]$ (mval/1 IZ)	$244,81 \pm 13,9$ (21)	$198,87 \pm 10,5$ (28)* +++	$202,42 \pm 14,2$ (15)* ++	$206,92 \pm 11,3$ (15)* ++
Herzmuskulatur				
H_2O (Gew. %)	$77,51 \pm 0,27$ (17)	$77,18 \pm 0,61$ (16)		
E Z (l/kg Fr.-Gew.)	$0,2075 \pm 0,0034$ (17)	$0,1940 \pm 0,003$ (18)* ++	$0,2053 \pm 0,0056$ (15)	$0,204 \pm 0,0032$ (15)
$[Na_i^+]$ (mval/1 IZ)	$11,13 \pm 0,98$ (25)	$21,05 \pm 1,52$ (31)* +++	$14,89 \pm 1,21$ (15)** ++	$14,47 \pm 1,33$ (15)* +
$[K_i^+]$ (mval/1 IZ)	$126,05 \pm 1,33$ (25)	$127,59 \pm 1,69$ (31)	$118,84 \pm 1,40$ (15)*, ** +++	$118,37 \pm 1,31$ *,** ++ (15)
Skelettmuskulatur				
I Z (l/kg Fr.-Gew.)	$68,115 \pm 0,245$ (15)	$70,10 \pm 0,193$ (18)* +++		$69,98 \pm 0,231$ (10)* ++
E Z (l/kg Fr.-Gew.)	$7,82 \pm 0,314$ (15)	$6,47 \pm 0,286$ (18)* +++		$6,321 \pm 0,334$ (10)* +++
$[Na_i^+]$ (mval/1 IZ)	$11,01 \pm 2,45$ (15)	$13,24 \pm 3,06$ (18)		$17,13 \pm 2,82$ (10)* +
$[K_i^+]$ (mval/1 IZ)	$141,6 \pm 1,88$ (15)	$138,3 \pm 2,31$ (18)		$133,5 \pm 2,98$ (10)* +++

Tab. 1. Signifikanzberechnung gegen Gruppe I = *, Gruppe II = **.
Signifikanz auf 5 % Niveau = +, auf 1 % Niveau = +++, auf 0, 1 % Niveau = +++
Angaben in ( ) = Anzahl der Beobachtungen.
Elektrolyt- und Wassergehalt sowie Größe des Intra- und Extrazellulärraumes
der Aorta sowie der Herz- und Skelettmuskulatur bei Kontrolltieren, Ratten
mit erblicher spontaner Hypertonie, von denen ein Teil mit Furosemid behandelt war, sowie bei Ratten mit DOCA-Hochdruck.

Anpassung der Extremitätendurchblutung an rhythmische Muskelarbeit und ihre Beziehungen zu Gasspannungen und Substratkonzentrationen im Blut

Von K. Caesar, D. Jeschke, H.-F. v. Oldershausen
Medizinische Universitäts-Klinik, Tübingen, Germany

Das tägliche Training von Muskelgruppen nur einer Extremität mit rhythmischen Kontraktionen erschien geeignet, Anpassungsvorgänge der Muskeldurchblutung und des Stoffwechsels an eine gesteigerte Leistungsforderung zu untersuchen. 10 gesunde, jugendliche Versuchspersonen führten während 12 Wochen tägliche Übungen der Unterarmmuskulatur rechts durch. Dazu mußte ein Gewicht von 2 kg an den Fingern durch Beugen und Strecken im Handgelenk 60 Mal in 2 Minuten angehoben werden. Vor und in zweiwöchigen Abständen während der Übungszeit wurden jeweils in Ruhe, sofort nach dieser Arbeitsbelastung und in der Erholungsphase mit einem Venenverschlußplethysmographen die Unterarmdurchblutung registriert. Laktat, Pyruvat, P_{O2} und P_{CO2} im Armvenenblut und in der Arteria femoralis bestimmten wir vor und nach 6 und 12 Wochen Training. Bei diesen standardisierten Arbeitsversuchen zur Durchblutungsmessung und Stoffwechseluntersuchung wurde in gleicher Weise wie beim täglichen Training verfahren, jedoch nur die Beugemuskulatur des Unterarmes belastet.

Die Ruhedurchblutung (Abb. 1) zeigt leichte Schwankungen mit geringer Tendenz zur Zunahme nach 10 und 12 Wochen Trainingszeit. Signifikante Veränderungen entstehen nicht. Zur Erfassung der Leistungsbreite der Extremitätendurchblutung wurde eine standardisierte Belastungsprüfung der Unterarmmuskulatur durchgeführt. Außer dem Sofortwert nach Belastung wurden wei-

Die Untersuchungen wurden durchgeführt mit Unterstützung durch das Kuratorium für Sportmedizinische Forschung.

tere Messungen der abklingenden Arbeitshyperämie 1, 3, 5 und 10 Minuten nach Belastungsende durchgeführt. Die maximale Arbeitsdurchblutung bleibt gleich groß während der Trainingszeit, die Rückbildung der Mehrdurchblutung erfolgt jedoch schneller als vor Trainingsbeginn. Hierdurch kommt es zu einer Veränderung im Profil der abklingenden Arbeitshyperämie im Unterarm (Abb. 1) nach der von uns gewählten zweiminütigen Arbeitsbelastung.

Diese raschere Rückbildung bei gleich großer Arbeitsleistung war statistisch zu belegen, indem am Ende der Versuchszeit der 10 Minutenwert um 2,60 $\pm$ 3,42 ml/100 ml/min signifikant erniedrigt war gegenüber dem Vergleichswert vor Trainingsbeginn (P =< 0,05).

Eine Adaptation des Zellstoffwechsels unter Muskeltraining kann sowohl mit einer verbesserten aeroben als auch anaeroben Energiegewinnung erfolgen, wie die zusammenfassende Darstellung der zahlreichen Untersuchungen von KEUL, DOLL und KEPPLER (1969) zeigt. Der mehrfach beschriebenen Möglichkeit des trainierten Muskels zur vermehrten Sauerstoffextraktion kommt bei den kurzfristigen Belastungsformen unserer Versuche offenbar keine große Bedeutung zu. Der Abfall der venösen und der gleichzeitige Anstieg der arteriellen O_2-Werte nach zweiminütiger Unterarmarbeit bleiben während des Muskeltrainings unverändert. Ein gesteigerter Energieumsatz, bzw. eine Umstellung auf eine mehr nutritive Muskeldurchblutung, kann sich in der vermehrten CO_2-Abgabe in das venöse Armblut nach Training bemerkbar machen. Bei unveränderten arteriellen Werten haben wir nach 6 und signifikant nach 12 Wochen Armtraining sofort nach Belastung höhere venöse P_{CO2}-Werte gemessen (P =< 0,05).

Die Belastungswerte für die venösen Laktatspiegel sind bei weitgehend unveränderten arteriellen Werten nach 6 und 12 Wochen der Versuchsdauer stärker und länger erhöht als vor Trainingsbeginn. Im Verhalten der Pyruvatspiegel wurden keine gerichteten Veränderungen beobachtet. Das zeigt sich im Laktat-Pyruvat-Quotienten, der entsprechend den Befunden für Laktat ebenfalls nach 6 und nach 12 Wochen Unterarmtraining beim Belastungsversuch höher ansteigt als vor Beginn des Übungsprogrammes. Als Ursache dieser vermehrten Laktatabgabe bei gleicher Arbeitsleistung nehmen wir eine erhöhte Glykolysekapazität als Trainingseffekt auf die Muskelzelle an.

Zusammenfassung: Als Anpassung an tägliche Muskelübungen mit kurzfristiger Belastung haben wir eine raschere Rückbildung der Arbeitshyperämie zu den Ruhewerten beobachtet. Aus den gleichzeitigen Befunden einer vermehrten Abgabe von Kohlendioxyd und Laktat schließen wir auf eine verbesserte aerobe und vor allem anaerobe Energiegewinnung der Muskelzelle. Weiterhin sehen wir in unserem Ergebnis einen Hinweis dafür, daß bei verminderten Durchströmungswerten nach Muskeltraining trotz erhöhten Laktatspiegeln und P_{CO2}-Drucken diese Stoffwechselfaktoren nicht für die effektivere Steuerung der peripheren Zirkulation maßgebend sein können.

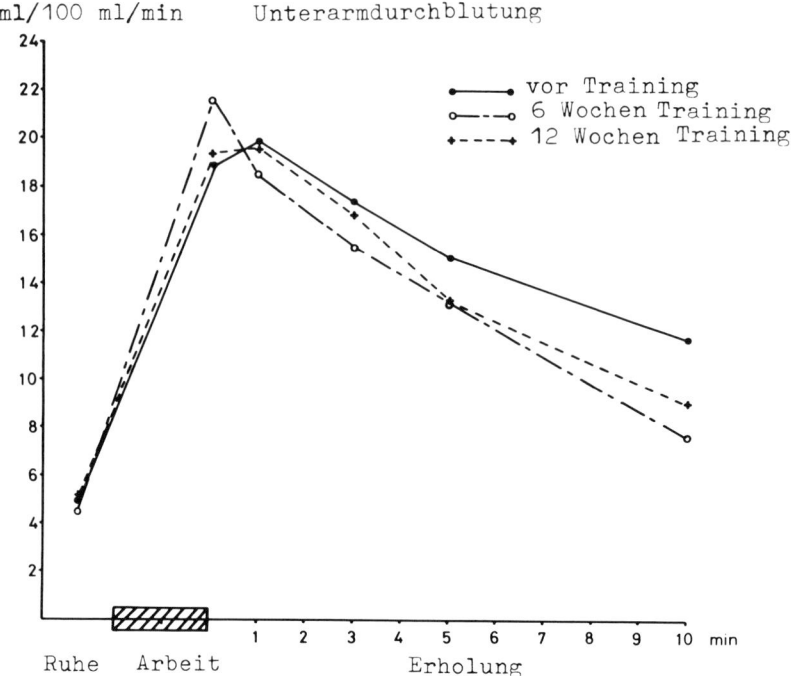

ml/100 ml/min   Unterarmdurchblutung

vor Training
6 Wochen Training
12 Wochen Training

Ruhe   Arbeit   Erholung

Abb. 1. Auswirkung täglicher Unterarmkontraktionen auf Ruhedurchblutung und Arbeitshyperämie im Unterarm. Mittelwerte aus 10 Versuchen

Literatur

1) ASTRAND, P.O., RODAHL, K.: Textbook of Work Physiology. New York: Mc Graw-Hill Book Company 1970.
2) KEUL, J., DOLL, E., KEPPLER, D.: Muskelstoffwechsel. München: Barth 1969.

Mechanism of Action of Hypoxia in Tracheal Smooth Muscle (TSM) with a Note on the Role of the Series Elastic Component

By N. L. Stephens
Department of Physiology, Faculty of Medicine, University of Manitoba,
Winnipeg, Canada

We have previously shown that hypoxia (P_{O2} 60 mmHg, P_{CO2} 40 mmHg and pH 7.40) produced a 30 % reduction of isometric tetanic tension (P_0), a 40 % reduction of rate of development of P_0, dP/dt, and a 15 % reduction of time to attain P_0 in canine TSM (1).

Concomitant biochemical studies (2), revealed that in hypoxia creatine phosphate decreased from 0.935 ± 0.069 (SE) to 0.16 ± 0.036 μmoles/gram wet weight and ATP decreased from 1.119 ± 0.042 (SE) to 0.681 ± 0.089 μmoles/ gram wet weight (P< 0.05). The ADP content (about 0.32 moles/g) was not significantly altered. The increase in AMP content from 0.048 ± 0.037 (SE) to 0.205 ± 0.037 μmoles/gram wet weight however was significant (P< 0.05).

These mechanical and biochemical studies demonstrate that changes in mechanical function of TSM are associated with measurable changes in levels of energy stores.

Assuming that the striated muscle model can be applied to smooth muscle we speculated firstly that the reduction in P_0 could be due to reduction in the number of forcegenerating actomyosin sites in the muscle, secondly that if one assumed the properties of the series elastic component (SEC) of the TSM were unchanged by hypoxia, the reduction in dP/dt could be due to reduced rates of energy utilization, and thirdly that the reduction in time to attain P_0 could be due to a quicker withdrawal of calcium from the troponin-tropomyosin system.

In this communication the results of biophysical experiments designed to determine whether reduced rates of energy utilization were responsible for the reduced dP/dt, are presented.

Employing A.V. HILL's hypothesis that the $\underline{b}$ constant, which has the units of velocity and which is derived from the force-velocity hyperbola of muscle, is an index of the rate at which energy utilizing reactions occur, we conducted experiments to determine if the value of $\underline{b}$ was reduced in hypoxia. We have previously shown that valid force-velocity curves which are hyperbolic in shape can be obtained from tracheal smooth muscle (3).

Conventional force-velocity curves were obtained for TSM during normoxia and during hypoxia (P_{O2} 40 mmHg, P_{CO2} 40 mmHg, pH 7.40) of an hours duration.

Supported by grants from the Medical Research Council of Canada and the Canadian Heart Foundation.

In fig. 1, load P is plotted along the abscissa in grams, P_0 representing the maximum load the muscle is just able to lift. In the left hand ordinate velocity is plotted in centimeters per second. The right hand ordinate is plotted in $(P_0-P)/\underline{v}$ units. This is used to obtain linear transform of a hyperbolic curve

The hyperbolic nature of the normoxic curve is evident and the parallel leftward shift of the hypoxic curve is striking. The straight lines were obtained by plotting $(P_0-P)/\underline{v}$ versus P. The two lines (slopes = $1/\underline{b}$ and intercepts = $\underline{a}/\underline{b}$) enabled computation of the conventional force-velocity constants which are shown in the insert in normalized units, A representing cross-sectional area of muscle and, L_0 standard muscle length. Reductions in both $\underline{a}$ and $\underline{b}$ are seen. These are statistically significant. A reduction in V_{max} (which is equal to $P_0 \underline{b}/\underline{a}$), the derived maximum velocity of shortening under a theoretical zero load, is also evident. The means and standard errors obtained from 8 similar experiments comparing the effects of normoxia and hypoxia are as follows: $\underline{a}/A$ (in g/cm^2) changed from 212 ± 32 (SE) to 70 ± 21; $\underline{b}$ (in L_0/sec) changed from 0.06 ± 0.006 (SE) to 0.02 ± 0.005; V_{max} (in L_0/sec) changed from 0.30 ± 0.02 (SE) to 0.20 ± 0.04, and P_0/A (g/cm^2) changed from 925 ± 77 (SE) to 308 ± 28.

With reference to the hypothesis under test the important finding is that $\underline{b}$ is significantly reduced from $0.06 \ L_0/sec$ to $0.02 \ L_0/sec$. From this we concluded that the rate of energy utilization is adversely affected by hypoxia.

Though P_0 and $\underline{a}$ are both reduced, the ratio $\underline{a}/P_0$ was not significantly altered and hence the reduction in V_{max} must result from the reduction in $\underline{b}$.

The reduction in $\underline{a}$ suggested that either the number of actomyosin sites participating in force-generation were less, or that a smaller amount of force was generated per site, the total number of sites being unchanged.

Changes in the physical properties of the series elastic component (SEC) do not influence our conclusions since it is held constant in eliciting force-velocity curves.

However since the changes seen in dP/dt during an hypoxic isometric contraction could result from changes in the properties of the SEC, the latter was studied.

The dynamic modulus of elasticity, dP/dL, was computed by simultaneously measuring rates of changes in length (dL/dt) and in tension (dP/dt) during isotonic contractions according to a method we have already reported (4). The quotient of these two derivatives provided dP/dL; $dP/dL = (dP/dt)/(dL/dt)$. Integration of the dP/dL vs P curve yielded the length-tension curve for the SEC. These curves were exponential in both normoxia and hypoxia and qualitatively resembled those for skeletal and cardiac muscle. Means and standard errors computed from data for normoxia and hypoxia were as follows: dP/dL changed from 7.46 ± 0.54 to 14.52 ± 0.90 g/cm; the computed extension of the SEC under a load equal to P_0 changed from $9 \% \pm 1 \%$ to $4 \% \pm 0.4 \%$ of L_0.

155

These results demonstrate that the modulus of elasticity is increased under hypoxia. Similar finding have been reported by HENDERSON et al. for rat heart muscle (5). The reduced extension under a P_O load seen with hypoxia is a function of the reduced P_O of hypoxia. However calculation of the extension for a P_O equal to that of normoxia still yielded a total extension less than that in normoxia. These changes indicate that the reduced dP/dt seen in hypoxia can not be due to increased compliance of the SEC, and must result from the decreased rates of energy utilization.

In conclusion we have shown that hypoxia reduced both the maximum active tension airway smooth muscle can develop and the velocity of isotonic shortening. It also reduced levels of creatine phosphate and adenosine triphosphate in the tissue.

Reduction in the <u>b</u> constant of the force-velocity equation of A.V. HILL's suggests that the reduction in shortening velocity is due to reduced rates of those reactions (hydrolysis of adenosine triphosphate and creatine phosphate being an important one), which provide energy for utilization during contraction. Since our previous reports indicated that membrane excitability was not altered (6) and intracellular calcium content was not decreased in hypoxia (2), the reduced rate of contraction may stem from changes in those reactions that are involved in energy utilization, and which follow successful excitation-contraction coupling.

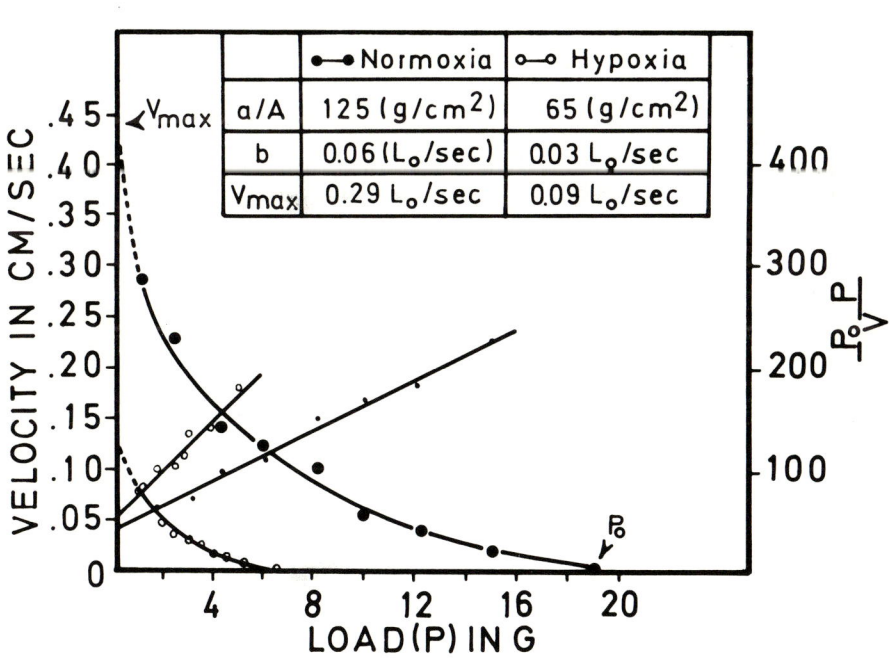

References

1) STEPHENS, N.L., CHIU, B.S.: Mechanical properties of tracheal smooth muscle and effects of O_2, CO_2 and pH. Amer. J. Physiol. 219, 1001-1008 (1970).

2) KROEGER, E., STEPHENS, N.L.: Effect of hypoxia on energy and calcium metabolism in airway smooth muscle. Amer. J. Physiol. 220, 1199-1204 (1971).

3) STEPHENS, N.L., KROEGER, E., MEHTA, J.A.: Force-velocity characteristics of respiratory airway smooth muscle. J. appl. Physiol. 26, 685-692 (1969).

4) STEPHENS, N.L., KROMER, U.: Series elastic component of tracheal smooth muscle. Amer. J. Physiol. 220 (1971) (in press).

5) HENDERSON, A.H., PARMLEY, W.W., SONNENBLICK, E.H.: The series elasticity of heart muscle during hypoxia. Cardiovasc. Res. 5, 10-14 (1971).

6) STEPHENS, N.L., KROEGER, E.: Effect of hypoxia on airway smooth muscle mechanics and electrophysiology. J. appl. Physiol. 28, 630-635 (1970).

Lokalisationsfaktoren bei Durchblutungsschäden

Von P. Sunder-Plassmann
Chirurgische Klinik der Universität Münster/Westf., Germany

Klinische und angiographische Beobachtungen plötzlicher Arterienverschlüsse peripherer (drohende Finger- und Handgangrän) und zentraler (A. cerebri media mit Hemiplegie und Aphasie) Art bei Kindern mit Penicillin- und Streptomycin-Allergie waren Veranlassung zu Vergleichen mit Gefäßprozessen be transplantierten Nieren auf immunologischer Grundlage. Seit 1965 führen wi in Münster zusammen mit der Internen Arbeitsgruppe um Prof. LOSSE Nierentransplantationen durch und hatten daher auch Gelegenheit, Gefäßveränderungen bei der 'homograft rejection' zusammen mit unseren Pathologen zu diskutieren.

Darunter war ein Fall bei einer 34-jährigen Frau besonders beachtlich: Es ging ihr nach der Nierentransplantation zunächst gut, die Anastomosen funktionierten sehr gut und die Niere schied auch gut aus. Aber am 13.-14. Tag setzten trotz üblicher immunsuppressiver Medikation Abstoßungszeichen ein die am 16. Tag an der Niere zu frischen Intima-Verkalkungen im Bereich vo

Schaumzell-Bildungen an größeren und kleineren Interlobär-Arterien, ausgeprägtem Media-Ödem der glatten Muskulatur und adventitiellen Lymphozyten-Infiltraten führten. Wir erhielten diese Niere damals zur Transplantation bei gutem 'tissue typing' in Zusammenarbeit mit 'Eurotransplant' in Leiden durch Prof. van ROOD, und da die zweite Niere jenes Spenders, die anderweitig transplantiert wurde, solche Reaktionen nicht zeigt und gut funktioniert, sind wir der Meinung, daß die immunologische Situation dieser 34-jährigen Frau mit starker individueller, also gen-fixierter Komponente, den Boden für jene eindrucksvollen Veränderungen abgab. Die bei ihr in der zweiten Woche nach der homologen Transplantation festgestellte Hyper-Calcaemie (Serum-Ca: 6,6 mvl) hatte außerdem zu einer Lungen-Calcinose geführt, die nach Ansicht der Pathologen zur Todesursache wurde.

Die geschilderten klinischen, angiographischen und mikroskopischen Befunde veranlaßten uns, einige hundert Tierversuche, über die wir schon vor 20 Jahren berichtet hatten, zusammen mit den Professoren MENGES (Anaesthesiologie), BACKMANN (Neurochirurgie) und SCHNEPPER (Radiologie) wieder aufzunehmen.

Besonders bei sensibilisierten Kaninchen (inakt. steriles Schweineserum) ließen sich innerhalb 2-6 Wochen durch Resektion, ja selbst durch ganz kurzfristige (1 Minute), umschriebene Quetschung vorgelagerter Parasympathikus-Segmente Gefäßschäden mit Intima-Ödem, Endothel-Ulcus, appositioneller Thrombenbildung, Elastika-Fragmentierung und Verkalkung in die jeweils abhängigen Gefäßbezirke z.B. die thorakale Aortenwand bei Eingriffen im zervikalen Vagusbereich lokalisieren, während hochdosierte Cholesterin-Fütterung bei Kontrollen im gleichen Zeitraum noch überhaupt keine Veränderungen der Aortenwand erkennen ließen.

An den Schilddrüsen jener vorbehandelten Tiere fand sich eine ausgeprägte Kolloidstapelung, im Sinne einer strumaähnlichen Anstauung, bei gleichzeitig sehr abgeflachtem Follikelepithel; und gleichzeitig ergab sich ein deutlicher Schwund der parafollikulären C-Zellen, denen bekanntlich die Calcitonin-Produktion zugeschrieben wird.

Wurde dieser depressorische Effekt auf die Schilddrüse noch durch Thio-Uracil potenziert, zeigten sich in solchen Serien geradezu groteske Ausmaße einer schwersten Aortenwand-Verkalkung, die mit gleichzeitiger Schaumzellen-Bildung und Elastika-Fragmentierung durchaus an menschliche Arteriosklerose erinnert.

Antikoagulantienprophylaxe in der Therapie chronischer Becken- und Beinarterienverschlüsse

Von H. D. Bruhn, P. Jipp, J. Schellmann, I. Sedlmeyer, H. Müller-Wiefel und D. Borm

Medizinische Klinik und Chirurgische Klinik der Universität Kiel, Germany

Die Studie umfaßt 198 stationäre Patienten der I. Medizinischen Klinik und der Chirurgischen Klinik der Universität Kiel mit chronischen Verschlüssen der Becken- und Beinarterien. In 333 Extremitätenarterien der 198 Patienten liegen arteriographisch gesicherte haemodynamisch signifikante arteriosklerotische Veränderungen in den verschiedenen Segmenten vor, wobei in der Mehrzahl der Fälle eine Durchblutungsinsuffizienz vom Grad II nach Fontaine besteht. Von den 198 Kranken konnten 158 nach im Mittel 19,3 Monaten nachuntersucht werden. Hiervon waren 63 Patienten konservativ und 95 Patienten chirurgisch (offene oder halbgeschlossene Thrombendarteriektomie von 130 Segmentverschlüssen an 116 Extremitäten) behandelt worden. Sowohl bei den 63 konservativ behandelten als auch bei den 95 operativ versorgten Kranken wurde der Therapieerfolg an der Änderung des Pulsstatus und des Grades der Durchblutungsinsuffizienz gemessen. Bei einer Verschlechterung des Pulsstatus erfolgte die Wiederaufnahme zur Kontrollangiographie. Der angestrebte optimale Bereich der Dicumarolprophylaxe lag zwischen 15 und 30 % Quickwert. Bei der Auswertung wurde ein Quotient gebildet, dessen Zähler die Häufigkeit der im optimalen Bereich liegenden Quickwerte wiedergab, während im Nenner die Zahl der über 30 % Quickwert liegenden Bestimmungen erschien. Bei einem Quotienten unter 10 nahmen wir eine schlechte, bei einem Quotienten von 10 und darüber eine gute Antikoagulation an.

Bei den konservativ behandelten Patienten dieser Studie wirkte sich die Dicumarolprophylaxe insofern günstig aus, als die an insgesamt 84 Extremitäten zu Beginn der Untersuchung palpierten 223 Pulse bei Kontrolle unverändert nachweisbar waren. Hiernach schien die Verschlußkrankheit nicht progredient gewesen zu sein.

Die Auswertung der chirurgisch behandelten Verschlußkranken unter dem Gesichtspunkt der Antikoagulantienprophylaxe ohne Berücksichtigung der Verschlußlokalisation zeigt folgendes Bild (Tab. 1): Bei guter Antikoagulation sind von insgesamt 68 erfolgreich operierten Verschlüssen bei der Nachuntersuchung 8 (12 %) wieder verschlossen. Bei schlechter Antikoagulation beträgt die Verschlußrate 20 % (9 von 46) und ohne Antikoagulation 37 % (6 von 16). Stellt man die Zahlen der postoperativ durchgängigen Segmente mit und ohne Antikoagulantienprophylaxe einander gegenüber, so ist der Unterschied statistisch signifikant (P< 0,05).

Die Frage nach der Wirkung der Antikoagulantienprophylaxe in Bezug auf die Verschlußlokalisation läßt sich nach dem vorliegenden Material nur indirekt beantworten. Es zeigt sich nämlich, daß die durchgängig gebliebenen Segmente im Femoropoplitealbereich in 61 % (37 von 61) gut antikoaguliert waren, im Beckenbereich in 51 % (19 von 37) und im Aortenbereich nur in 44 % (4 von 9). Im Femoropoplitealbereich ist also bei erhaltenem Operationsresultat der Anteil der gut antikoagulierten Segmente prozentual am höchsten. Daraus kann die Tendenz abgeleitet werden, daß die Langzeitantikoagulation im Femoropoplitealbereich zur Aufrechterhaltung des Operationsergebnisses von relativ größerer Bedeutung ist, als in anderen Strombahnanteilen. Dieser Befund wird noch dadurch gestützt, daß die Femoropoplitealregion mit 24 % die höchste Rethrombosierungsrate aufweist, während die Beckenstrombahn nur mit einer Rethrombosierungsrate von 10 % belastet ist. Die 9 wiedereröffneten Aortensegmente blieben sämtlich durchgängig.

Aus den vorgelegten Befunden ergibt sich eine klare Indikation zur Antikoagulantienprophylaxe nach rekonstruktiven Gefäßoperationen. Die Langzeitprophylaxe sollte, falls auch im weiteren Verlauf keine Kontraindikationen auftreten, lebenslang fortgesetzt werden, da die Disposition zur Verschlußkrankheit unverändert weiter besteht.

| | Antikoagulantien-Einstellung | | |
	gut	schlecht	keine
Segmentzahl	68	46	16
wieder-verschlossen	12 %	20 %	37 %

Summary and Discussion

The question of Graf KEYSERLINGK "what is the mechanism for the increased stiffness of the series elastic component in hypoxia" can not yet be answered. The effect of hypoxia on the contractile function appears within about 5 minutes (BERNE, STEPHENS). There is a rapid fall over the next 30 minutes and then a slow fall for the following 30 minutes. Is the mechanical response seen by STEPHENS on electrical stimulation the result of direct excitation of muscle (VANHOUTTE)?

STEPHENS answered that the evoced response is nerve-mediated. The muscle is of the multi-unit type; its neuro-effector is Acetylcholine. On blocking with Atropine the mechanical response is reduced to about 25 %. Considerably stronger stimuli can elicit the full response again but the preparation does not stay stable very long.

BALOURDAS:
Has the hypoxia in congestive heart failure with disturbances of energy utilization-mechanism and reduction of cardiac efficiency any relationship to your findings and to your experiments on tracheal smooth muscle?

STEPHENS answered: Yes, I think our findings simulate fairly closely the findings in hypoxic (ischaemic) heart muscle. We observe considerable reduction in CP with only a very small reduction in ATP levels. The answer given for Dr. RANGACHARIS question applies here also.

When asked whether he has any experimental evidence as to changes in the calcium binding of troponin in his preparations he answered that HONG et al. have commenced looking at regulatory properties in smooth muscle and he thinks to utilize their technique to determine if hypoxia impairs the calcium binding or releasing properties of troponin.

Chairman: T. A. Balourdas

Myocardial Mechanical, Ionic, Metabolic and Electrical Correlates in Response to Coronary Ischemia

By J. P. Gilmore, R. Jacob, J. Nizolek, and D. T. Miller
Department of Physiology, University of Virginia and Department of Physiology
and Biophysics, University of Nebraska Medical Center, USA

The experiments to be presented were carried out employing the blood perfused dog Langendorff preparation. Myocardial performance was monitored
using a strain gauge arch sewn to the wall of the ventricle with the preload
set at about 130 % of resting fiber length. Blood was diverted from the coronary inflow and coronary outflow lines and passed through a Guyton A-VO$_2$
analyzer to continuously monitor coronary arterio-venous O$_2$ difference and
the blood then returned to the dog which perfused the isolated heart. The
A-VO$_2$ difference of the preparation at normal coronary blood flow was approximately 7 to 8 volumes percent. This along with coronary blood flow provided us with a continuous measurement of myocardial oxygen consumption.
Myocardial arterio-venous blood potassium difference was continuously monitored using an on-line automated flame photometery system. To obtain measurements of the myocardial transmembrane potential, the technique of PRU-
ETT and WOOD was employed. This is a technique in which a micromanipu-
lator is mounted on a small plastic disc and the disc in turn sewn to the wall
of the ventricle. Under any given experimental condition several myocardial
cells were penetrated for potential measurements. The potentials were analyzed primarily in terms of the first derivative and the duration of the action
potential plateau. Unless indicated otherwise heart rate was maintained constant through external pacing. Coronary perfusion pressure was altered by a
clamp on the coronary inflow line.

The initial studies were concerned with the effects of altering coronary perfusion pressure on myocardial potassium balance, oxygen consumption and
performance. We found a linear relationship between coronary perfusion pressure and blood flow, i. e. there was no consistent evidence that autoregulation
of coronary blood flow occurred. When blood flow was reduced greater than
approximately 60 %, coronary venous blood potassium concentration increased. The relationship between myocardial blood flow and myocardial oxygen
consumption was curvilinear, in that the decrease in oxygen consumption was,
on the average, considerably less than the decrease in coronary blood flow.
For example, a 30 % decrease in coronary blood flow was associated with
approximately a 10 % decrease in myocardial oxygen consumption. It was
only under conditions in which coronary venous blood potassium increased
indicating true ischemia that reductions in coronary blood flow produced relatively proportional reductions in myocardial oxygen consumption. Also, when
there was no evidence of ischemia large changes in coronary blood flow produced only small changes in ventricular performance. However, the change
in ventricular force was proportional to the change in coronary blood flow
when evidence of myocardial ischemia obtained. In contrast, the correlation
between myocardial oxygen consumption and force was relatively linear.

These data show that reductions in coronary perfusion pressure and/or blood flow are associated with reductions in both myocardial performance and myocardial oxygen consumption. It has been suggested that the mechanism relating to this effect of blood flow on oxygen consumption is ischemia. For example, when coronary blood flow is reduced areas of the heart would become ischemic as a result of reduced oxygen delivery which in turn would reduce the oxygen consumption of the heart. We doubt that ischemia is the explanation for the alteration of myocardial oxygen consumption when blood flow is reduced except when coronary flow is severely limited. Our own experiments show that when large decreases in oxygen consumption were associated with decreases in coronary blood flow coronary venous blood potassium levels invariably increased, a reflection of cardiac ischemia. We are more inclined to believe that under non-ischemic conditions the reduction in oxygen consumption associated with reductions in coronary blood flow is the result rather than the cause of the associated decrease in ventricular contraction. Our experiments, of course, do not provide insight as to why myocardial contraction is decreased when perfusion pressure and flow are decreased. It is possible that it is related to what Dr. ARNOLD and associates have referred to as the "garden hose effect", that is, when perfusion pressure is decreased the stretch on the myocardial fiber is decreased so that the force of contraction is decreased. However, against this is our observation that when coronary perfusion pressure was decreased there was no consistent change in the level of resting diastolic tension as indicated by the strain gauge arch. The decrease in ventricular contraction caused by decreasing coronary blood flow could be due to the accumulation of a substance such as adenosine which has a negative inotropic effect on the heart. However, if significant amounts of adenosine accumulated when coronary blood flow is reduced, we would not have expected the linear relationship between coronary perfusion pressure and blood flow observed in our experiments. Our failure to find evidence of autoregulation in this preparation was of interest. Our average coronary blood flows were about 60-70 ml a minute and coronary perfusion pressure 100 mmHg, values that have been reported for the intact dog heart. However, because it is a nonworking preparation the oxygen consumption is low due to a decrease in the coronary A-VO$_2$ difference. Therefore, in this preparation coronary venous P$_{O2}$ is high. There are those who have suggested that coronary resistance is importantly controlled by intracellular P$_{O2}$. We would assume that since venous P$_{O2}$ was high intercellular P$_{O2}$ was also high, and yet, coronary resistance was normal. If P$_{O2}$ through some mechanism such as adenosine production is controlling coronary resistance we would have expected a high rather than normal coronary vascular resistance.

The results of our electrophysiologic studies show an association between the decrease in the force of cardiac contraction when coronary blood flow is greatly reduced and a closely coupled shortening of the myocardial action potential plateau. When these electrophysiologic changes occur there is invariably an increase in coronary venous blood potassium concentration. It has been reported that an increase in extracellular potassium decreases the duration of the action potential plateau and the force of myocardial contraction. Although the increase in coronary venous potassium concentration observed during ischemia was about 2 mEq/l, this does not mean that the extracellular potassium has increased only to this extent. One might expect a substantial gradient between the extracellular and intravascular space. Therefore, it is possible that extracellular potassium can rise during ischemia to such an extent that it contributes directly to the shortening of the plateau and thereby be closely related to the associated decrease in cardiac contraction. Although it might be suggested that the depression of performance observed under conditions of ischemia was due to lack of oxygen per se, against this is the finding of several others that under conditions of low oxygen cardiac muscle will still respond to positive inotropic interventions. In conclusion, therefore, reducing coronary blood flow to non-ischemic levels causes a depression of both the performance of the heart and its oxygen consumption. It appears that the decrease in oxygen consumption is a result rather than the cause of the associated decrease in contractile force. The decrease in myocardial performance observed when coronary blood flow is reduced to ischemic levels may be due to changes in the transmembrane potential of the cardiac muscle.

Discussion

Dr. BERNE discussed the meaning of Dr. GILMORE's results, particulary the high venous P_{O2} of his preparation, with respect to the mechanisms of autoregulation. Although he agreed with Dr. GILMORE that the mechanism for the decrease in cardiac performance seen when coronary blood flow is decreased is not due to the accumulation of adenosine, his reason for believing so, is that the amount of adenosine which would accumulate with reduction in coronary blood flow is not sufficient to produce a decrease in myocardial performance. It takes very large doses of adenosine to cause myocardial depression.

Dr. GILMORE agreed. His only point was that if adenosine does indeed accumulate with a reduction in coronary blood flow and is contributing to autoregulation of coronary blood flow, why autoregulation did not occur in his preparation when coronary perfusion pressure was varied over "a physiologic range"?

Dr. BETZ and Dr. LASSEN pointed out to the possibility of acid base changes and changes of K^+ distribution which might occur in these preparation during decreased perfusion pressure and might contribute to the observed effects. Dr. GILMORE admitted that acid base changes possibly contribute to the myo cardial depression observed when true ischemia is produced, i.e., when the reduction in coronary blood flow is associated with a myocardial loss of potassium. However, he would doubt very much if such changes played an important role over that range of coronary perfusion pressure whereby performance is decreased, but yet at which ischemia does not occur, i.e. there is no loss of myocardial potassium. Under these latter conditions coronary venous blood still has a very high oxygen content.

Dr. KOEPCHEN was interested in the mechanism whereby ischemia shortens the action potential and thus decreases myocardial performance. Dr. GILMORE's present position is that in some manner the high extracellular potassium decreases the inward calcium current thereby shortening the action potential which then decreases myocardial performance.

Chairman: R. Jacob

Die Bedeutung der Gefäßwandmuskelzellen für die Entstehung der Arteriosklerose. Elektronenmikroskopische und immunhistochemische Befunde

Von H.-J. Knieriem
Pathologisches Institut der Universität Düsseldorf, Germany

Elektronenmikroskopische Untersuchungen der Intimapolster bei der spontanen Arteriosklerose des Rindes und der experimentellen Atheromatose des Kaninchens zeigen, daß die Plaques durch proliferierte subendotheliale und intima-nahe, mediale glatte Muskelzellen entstehen. Diese glatten Muskelzellen der Gefäßwand können bei Hyperlipämie und Hypercholesterinämie zu großen Schaumzellen umgewandelt werden und dann atheromatöse Herde bilden. Die proliferierten Gefäßwandmuskelzellen sind durch ihren Reichtum an Mitochondrien, Ribosomen und durch ein erweitertes endoplasmatisches Retikulum gekennzeichnet und werden auch als "aktive" oder "modifizierte" glatte Muskelzellen bezeichnet. Die proliferierten glatten Muskelzellen sind durch spezifische, mit Fluoreszein markierte Antiseren gegen Human- oder Rinder-Myosin und Actomyosin identifiziert worden. Die immunhistochemischen Befunde wurden an normalen und arteriosklerotisch veränderten Gefäßen des Rindes und an atherosklerotischen Polstern der Aorta, den Koronararterien sowie den Pulmonal- und Hirnbasisarterien des Menschen erhoben und miteinander verglichen. Bei den arteriosklerotischen Herden des Menschen liessen sich in den proliferierten Muskelzellen neben Neutralfetten auch abgelagerte und gespeicherte Serum-Betalipoproteide durch ergänzende immunhistochemische Untersuchungen mit spezifischen, Fluoreszein markierten Antiseren nachweisen. Trotz verschiedener ätiologischer Faktoren (mechanische und hämodynamische Faktoren, Hyperlipidämie, Hypoxie und auch infektiös-toxische Noxen) kann beim Kaninchen, beim Rind und auch beim Menschen eine formal einheitliche Pathohistogenese der Arteriosklerose angenommen werden. Infolge der Pluripotenz der glatten Muskelzellen, die sowohl in der Lage sind neben den kontraktilen Proteinen Myosin und Actomyosin, auch Kollagen, elastisches Material und die Mucopolysaccharide der Grundsubstanz zu bilden, ist es verständlich, wenn die arteriosklerotischen Gefäßveränderungen unterschiedlich ausgeprägt sind. Neben atheromatösen, lipoidreichen Intimapolstern wie sie beim Menschen und experimentell beim Kaninchen vorkommen, finden wir lipoidarme, fibröse Polster und Plaques , die besonders reich an Kollagen und Mucopolysacchariden sind. Die morphologischen Untersuchungen belegen, daß bei der Arteriosklerose und auch bei der Koronarsklerose, die Zelle, die glatte Muskelzelle der Gefäßwand im Mittelpunkt der Prozesse steht, die zur Umformung, Verhärtung und Lichtungseinengung der Gefäße führen und schließlich die Durchblutungsnot der Organe bedingen. Die Proliferation der Gefäßwandmuskelzellen kann als ein reparativer Prozeß aufgefaßt werden, da wir ähnliche proliferative Veränderungen auch bei traumatischer Schädigung der Intima und bei der Organisation arterieller Thromben beobachtet haben.

Literatur

1) JURUKOVA, Z. und H.-J. KNIERIEM: Elektronenmikroskopische Untersuchungen über die Organisation arterieller Thromben. Virchows Arch. path. Anat. 349, 368 - 381 (1970)

2) KNIERIEM, H.-J.:Elektronenmikroskopische Untersuchungen zur Bedeutung der glatten Muskelzellen für die Pathohistogenese der Arteriosklerose. Beitr. path. Anat. 140, 298-332 (1970)

3) KNIERIEM, H.-J.:Immunhistochemische Untersuchungen zur Bedeutung der glatten Muskelzellen für die Pathohistogenese der Arteriosklerose des Menschen. Beitr. path. Anat. 141, 4 - 18 (1970)

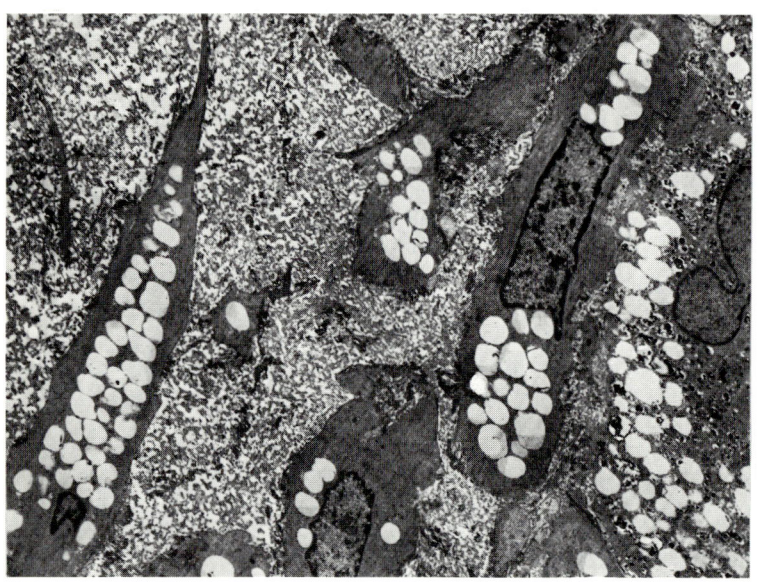

Abb. a: Cholesterin-Atheromatose des Kaninchens. Zahlreiche zu Schaumzellen umgewandelte proliferierte glatte Muskelzellen in einem großen Intimapolster. Vergr. 5000 x

Abb. b: Proliferierte glatte Muskelzellen in einem Intimapolster der Kaninchenaorta nach traumatischer Schädigung. Vergr. 16 000 x

Änderungen der Serum-LDH- und GOT-Isoenzyme während temporärer Myokardischämie

Von F. Z. Zàzvorka, E. Hoffmann, K. D. Rumpf, P. Walter, F. Grögler,
C. Beddermann, W. Lamprecht und H. G. Borst
Department für Biochemie und Department für Chirurgie der Medizinischen
Hochschule Hannover, Germany

Die Kenntnis der tolerablen Ischämiezeit ist die Voraussetzung der Behandlung des akuten Herzinfarktes durch einen aortakoronaren Bypass. Als Ziel dieser Untersuchungen wurde die Änderung des LDH-Isoenzymmusters und das Auftreten des mitochondrialen GOT-Isoenzyms in Abhängigkeit von der Ischämiezeit bestimmt. Bei 6 Schweinen wurde ein Myocardinfarkt durch Koronarokklusion für 3 Std. erzeugt und anschließend durch Entfernung der Gefäßklemme der Blutfluß für weitere 2 Std. wieder hergestellt.. Zur Bestimmung der LDH- und GOT-Isoenzyme wurde eine Koronarvene im Infarktgebiet kanüliert und Blut aus dieser und einer peripheren Vene entnommen. Serum-LDH- und GOT-Isoenzyme wurden mittels Agar-Elektrophorese bestimmt. Der Ablauf der Veränderungen der Serum-LDH- und GOT-Isoenzymaktivitäten ist in den Abbildungen zusammengefaßt. Unsere Resultate stimmen mit denen von SATO et al. veröffentlichten überein, die eine wesentliche Steigerung der LDH-Aktivität im Koronarblut bei Hunden nach 3o-minütiger Okklusion gefunden haben. Eine Erhöhung der LDH-Aktivität in Perfusaten der isolierten Rattenherzen hat de LEIRIS et al. nach 1o- bis 15-minütiger Hypoxie, die durch eine mit N_2 saturierte Lösung induziert wurde, festgestellt. GUDBJARNASON, SATO und auch wir haben in früheren Versuchsserien eine Abnahme der myokardialen LDH-Isoenzyme 1 und 2 und eine relative Zunahme der LDH-3, 4 und 5 in Hundeherzen nach 6, 24 und 48 Std. Okklusion festgestellt.

Der durch die Ischämie geänderte Substratfluß in den Zellen führt nach einem sich neubildenden metabolischen Status zur Einstellung solcher Substratspiegel, die zumeist unterhalb oder oberhalb der normalerweise erwarteten Sättigungskonzentration liegen. Die Enzymaktivitäten in den Zellen unterliegen dann nicht mehr den physiologischen regulatorischen Mechanismen. Der Stoff- und Substratfluß erfolgt nicht nur einseitig, d. h. eliminierend. Man muß annehmen, daß die geschädigte Membran auch den Eintritt verschiedener Stoffe erlaubt, wobei es zu einer Reihe von Inhibitions- und Konkurrenzerscheinungen zwischen Substraten und Enzymen kommen kann, die wir als kinetische Modulation bezeichnen. Derartige Effekte beeinflussen auch die Aktivitäten der LDH-Isoenzyme; so vor allem Substanzen, die im Zellstoffwechsel als Effektoren die Aktivitätsveränderungen der LDH steuern können, wie Oxalacetat, Pyruvat u. a.

Der akut einsetzenden Hypoxie der Zelle folgt unmittelbar ein hoher Anstieg des Lactats, wie sich im abführenden Venenblut zeigt. Dort ist außerdem ein gleichsinniger, sofortiger Pyruvatanstieg zu finden. Durch die anoxisch erhöhte Glykolyserate entsteht ein hoher Anfall an Pyruvat, der die kinetische Modulation der LDH auszulösen scheint. Hierbei steigt der LDH-5-Anteil relativ an.

Das kathodische GOT-Isoenzym erscheint im Koronarvenenblut nach 2o-3o Minuten, in der Peripherie ist es jedoch erst nach 4o-6o Minuten nachweisbar. Der Anteil des kathodischen Isoenzyms steigt im Koronarvenenblut bis zum Ende der Okklusion, nach der Revaskularisation nimmt die Aktivität rasch ab. In der Peripherie nimmt die Steigerung der Aktivität viel langsamer zu, dauert jedoch trotz Revaskularisation noch weiter an.

Nach Wiedereröffnung des Koronarflusses erfolgt neben einer Stoffwechselumstellung der noch nicht irreversibel geschädigten Zellen auf aerobe Verhältnisse ein großes Substratangebot von außen. Dadurch erklärt sich möglicherweise eine zunächst gleichsinnige Steigerung von Lactat und Pyruvat. Nach einer Latenzzeit von ca. 3o-45 Minuten setzt ein Abfall von Substraten in den Normalwerten ein. Gleichzeitig prägt sich auch eine Verschiebung im Isoenzymmuster wieder zugunsten der LDH-1 aus.

Hieraus läßt sich schließen, daß nach 3-stündiger Koronarokklusion ein Teil der Zellen des Infarktareales noch in der Lage ist, im Sinne einer Stoffwechselregulation bzw. Gegenregulation biologisch aktiv tätig zu sein. Dabei hat sich das kathodische Isoenzym der GOT als ein empfindlicher Indikator der groben Schädigung subzellulärer Strukturen gezeigt.

Literatur

GUDBJARNASON, S. , C. de SCHRYVER, C. CHIBA, J. YANAMAKA, R. J. BING: Protein and nucleic acid synthesis during the reparative processes following myocardial infarction. Circulat. Res. 15, 32o (1964)

SATO, H. : On alterations of myocardial lactic dehydrogenase isoenzyme by coronary ligation. Jap. Circulat. J. (En.) 33, 7o1 (1969)

WALTER, P. , F. Z. ZÀZVORKA, H. G. BORST, W. LAMPRECHT:Alterations of LDH-isoenzymes in infarcted heart muscle after transmural punction. European Surgical Research (in press).

ZÀZVORKA, F. Z. , J. KAMARÝT: LDH-, MDH- und GOT-Isoenzyme in der Frühdiagnostik der Leber-, Herz- und Bluterkrankungen. Ergebnisse der Laboratoriumsmedizin, Medicus Verlag Berlin 2, 41 (1965)

ZÀZVORKA, F.Z. , P.WALTER, F.GRÖGLER, C.BEDDERMANN,H.G.BORST and W. LAMPRECHT: Lactate dehydrogenase isoenzyme pattern in prolonged infarction and subsequent restoration of coronary blood flow. Gemeinsame Arbeitstagung Klinische Chemie, 22. -24. 4. 1961 in Wien

de LEIRIS, J. et al. : Arch. int. Physiol. 77, 749 (1969)

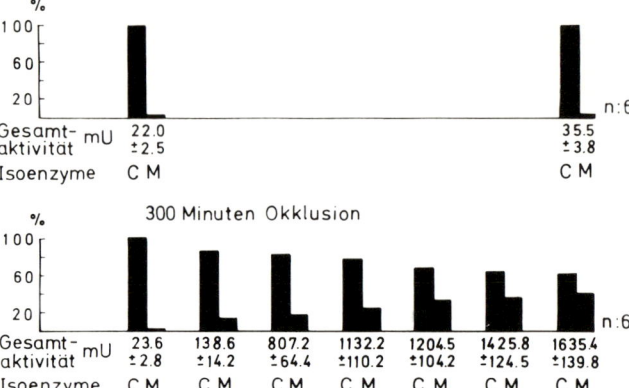

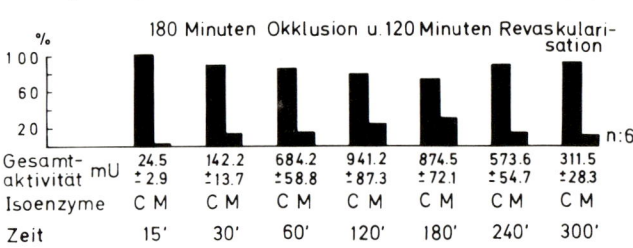

Abb. 1. Änderungen der Serum-GOT-Isoenzyme während temporärer Myocard-ischämie

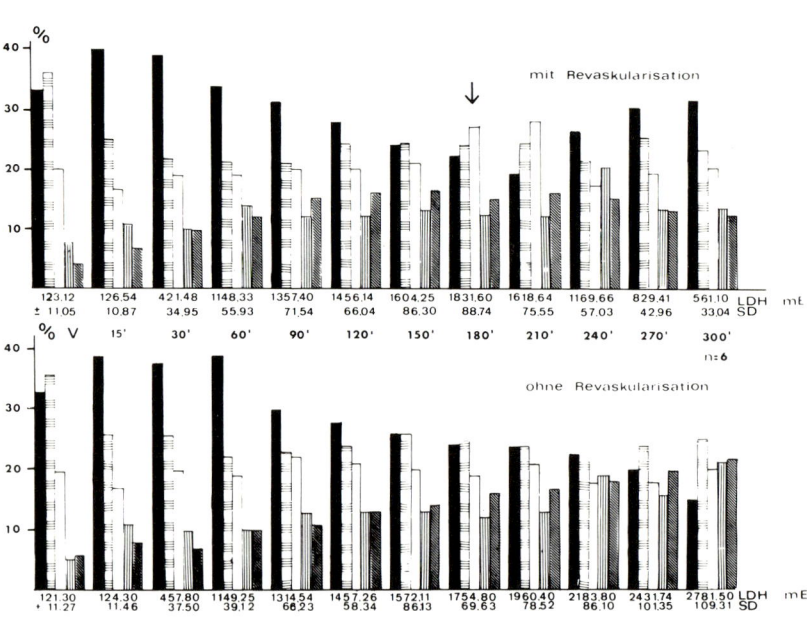

Abb. 2. Verlauf der Serum-LDH-Isoenzyme während temporärer Myocard-Ischämie im apikalen Ast der Vena magna cordis

Hemmung arteriosklerotischer Gefäßprozesse durch prophylaktische Behandlung mit MgCl$_2$ oder organischen Ca^{++}-Antagonisten. (Quantitative Studien mit Ca^{45} bei Ratten)

Von J. Janke, B. Hein, O. Pachinger, O. Leder und A. Fleckenstein
Physiologisches Institut der Universität Freiburg i. Br., Germany

Überdosierung von Vitamin D oder Dihydrotachysterol (DHT)führt insbesondere bei kombinierter Verabreichung mit NaH$_2$PO$_4$ bei Ratten nicht nur
zu Myokard-Läsionen sondern auch zu arteriosklerotischen Gefäßprozessen
(1). Nach unseren vorausgegangenen Studien mit markiertem ^{45}Ca ist die
Nekrotisierung durch eine Ca^{++}-Überladung der Myokardfasern bedingt und
daher nach Schwere und Ausdehnung mit der ^{45}Ca-Inkorporation quantitativ
korreliert (2, 3). Organische Ca^{++}-Antagonisten wie Verapamil, Substanz D
6oo oder Prenylamin, die den transmembranären Ca^{++}-Influx in die Myokardfasern reduzieren, konnten daher die Nekrosebildung verhüten. Ähnlich wirkten K$^+$- und Mg^{++}-Salze als natürliche Ca^{++}-Antagonisten. Ganz
analoge Verhältnisse wurden nunmehr an den Gefäßen gefunden: Ratten wurden 1o Tage lang peroral tägl. mit 2 x 1o mM/kg NaH$_2$PO$_4$ und 7 Tage lang
peroral mit tägl. 1 x o, 5 mg/kg DHT vorbehandelt. Am 11. Tag wurde intraperitoneal ^{45}Ca (1o /uC/kg)injiziert und die ^{45}Ca-Inkorporation in Aorta
und Art. mesenteric. sup. gemessen. Bei den unbehandelten Kontrollratten
erreichte der ^{45}Ca-Gehalt in der Aortenwand innerhalb von 6 Std. nach der
^{45}Ca-Injektion 7o % und in der Mesenterial-Arterie etwa 1oo %(bezogen
auf die Plasma-Aktivität = 1oo %), während bei den vorbehandelten Tieren
der ^{45}Ca-Gehalt der Aorta innerhalb der gleichen Beobachtungszeit auf
8ooo %, d. h. auf das 115-fache und der Mesenterial-Arterie auf 42oo %, d.
h. auf das 42-fache anstieg. Histologisch liessen sich dabei intensive
Media-Verkalkungen vom MÖNCKEBERG-Typ nachweisen. Gleichzeitig mit
DHT und NaH$_2$PO$_4$ verabreichtes MgCl$_2$ (2 x 7, 5 mM/kg tägl. peroral)konn-
te die excessive ^{45}Ca-Inkorporation in die Gefäßmuskulatur der Arterien
sowie die histologisch erfaßbaren Verkalkungen (ebenso wie die Myokardnekrosen) vollständig verhüten. Organische Ca^{++}-Antagonisten (Verapamil,
D 6oo) wirkten gleichsinnig.

Literatur

1)SELYE, H. : Amer. Heart J. <u>55</u>, 8o5 - 8o9 (1958)
2)FLECKENSTEIN, A. : Myokardstoffwechsel und Nekrose. In "Herzinfarkt
    und Schock" ed. von L. HEILMEYER und H. J. HOLTMEYER, G. Thieme,
    Stuttgart, 94 - 1o9 (1968)
3)JANKE, J. , A. FLECKENSTEIN und W. JAEDICKE:Pflügers Arch. <u>316</u>,
    R 1o (197o); <u>319</u>, R8, R 9 (197o).

The Effect of Local Ischemia on the Ionic Activity of Dog Myocardial Interstitium

By H. Benzing, M. Strohm, and G. Gebert
Physiologisches Institut I der Universität Tübingen und Abteilung Physiologie
der Universität Ulm, Germany

During the occlusion of a coronary artery only a small increase in coronary sinus blood H^+ and K^+ activity has been observed (2, 5). In order to study the real ionic changes in the interstitium of the ischemic myocardium
we measured continuously and simultaneously the changes of the H^+, Na^+
and K^+ activity by means of ion sensitive glass microelectrodes (see methods 3, 4, 1). A balloon constrictor for transient occlusion was placed
around the left descending coronary artery. Up to 7 glass microelectrodes
($\emptyset$ 5o-15oμ) were inserted into the region of the left ventricular wall which
was obviously supplied by the artery prepared for occlusion. Control electrodes were inserted into other parts of the myocardium. Furthermore we
recorded the local myocardial blood flow by means of heated thermo-
couples, the mean arterial blood pressure, the endexspiratory CO_2 concentration, the heart rate, and the ECG. We measured intermittently the arterial pO_2, pCO_2, pH, and the arterial Na^+ and K^+ concentration. The experiments were performed on 1o anesthetized, open-chest mongrel dogs
weighing 1o-2o kg.

1o-2o sec after occlusion the potentials of those electrodes changed which
were inserted in the ischemic region. Fig. 1 shows a short-lasting occlusion of the descending artery for about 1 min. The H^+ activity increased,
followed by a further increase after release of the blood flow (the increase
of the H^+ activity corresponds to a pH decrease of about o.15 pH units). The
increase of the H^+activity depended on the occlusion time, on the extent of
the ischemic region and on the localization of the electrodes. When the
occlusion time reached 15-3o min we found a pH decrease of about o, 65 pH
units (mean value) without additional increase of the H^+ activity after release of the occlusion. After heart fibrillation the H^+ activity increased
further in the ischemic region and started to increase in the non-ischemic
region both reaching a mean value of about 1.2o pH units (below the initial
value). The local myocardial blood flow which was diminished by occlusion
decreased to final values after fibrillation.

Besides the changes of the H^+ activity we found an increase in the K^+ activity and a decrease in the Na^+ activity in the myocardial interstitium
(fig. 1). In this experiment the K^+ activity increased from about 4 meq/l
(measured in the arterial blood) to about 22 meq/l and the Na^+ activity
decreased from about 15o meq/l(measured in the arterial blood)to about
1o5 meq/l. All the ionic changes were completely reversible after release
of the blood flow. H^+, Na^+ and K^+ electrodes inserted in normal supplied
regions showed no changes of the ionic activity.

It was only possible to measure changes of the ionic activity, and it was
impossible to get precise quantitative values of the K^+ shifts because of

the poor selectivity of the potassium electrodes (K^+ : Na^+ = 2-4 : 1). But it was possible to demonstrate the velocity and the extent of ionic activity changes in the interstitium of an ischemic region that could provide a basis for better understanding the behaviour of the ischemic myocardial cell.

References

1)BENZING, H., G.GEBERT, M.STROHM: Ionenverschiebungen im Myokard des narkotisierten Hundes unter verschiedenen Bedingungen. Verh.dtsch. Ges.Kreisl.-Forsch. (im Druck)
2)CRAMPTON, R.S., H.A.ROSELLE, and R.B.CASE: Coronary sinus blood pH during myocardial ischemia. Fed.Proc. 25, 4o2 (1966)
3)GEBERT, G.: Fortlaufende Messung der Aktivität von Alkaliionen im Extracellulärraum der Skeletmuskulatur. Habil.Schrift,Tübingen 1969.
4)GEBERT, G., H.BENZING, M.STROHM: Changes in the Interstitial pH of Dog Myocardium in Response to Local Ischemia, Hypoxia, Hyper- and Hypocapnia, Measured Continuously by Means of Glass Microelectrodes. Pflügers Arch. 329, 72 - 81 (1971)
5)KREUZIGER, H., H.ASTEROTH und K.ZIPF: Kaliumveränderungen nach Herzinfarkt im Tierexperiment. Z.Kreisl.-Forsch. 43, 382-387 (1954)

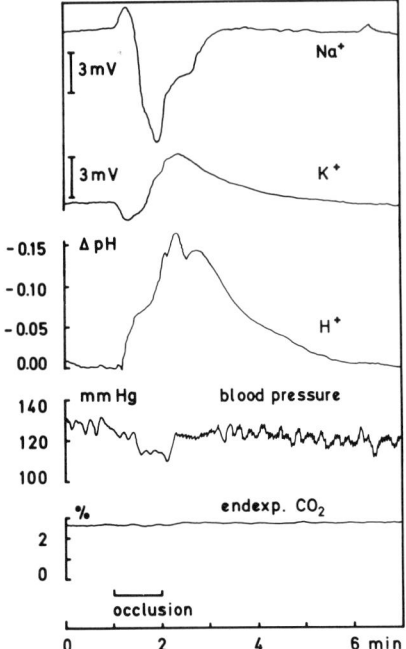

Effects of occlusion of the left descending artery for 1 min on (from top to bottom) the Na^+, K^+, H^+ activity, arterial blood pressure and exp. CO_2 concentration (maximum decrease of the K^+ activity about 45 meq/l; maximum increase of the K^+ activity about 18 meq/l; maximum decrease of the pH about o.15 pH units).

Regionale Myocarddurchblutung nach experimenteller Coronarocclusion

Von H. Flohr, N. Hahn, R. Felix, P. Lotz, D. Look und D. Redel
Physiologisches Institut der Universität Bonn, Germany

Die Durchblutung des Myocards unterliegt zeitlichen Schwankungen, die durch die mit der Kontraktion des Muskels einhergehenden Änderungen des Gewebsdruckes bedingt sind. Diese mit der Kontraktion auftretenden Drücke sind sowohl in ihrer Höhe als auch in ihrem Verlauf in den verschiedenen Regionen des Myocards verschieden.

Aufgrund theoretischer Überlegungen und der vorliegenden Messungen ist anzunehmen, daß der während der Kontraktion in der Wand des linken Ventrikels entwickelte Druck von außen nach innen zunimmt und in den subendocardialen Schichten den Aortendruck übersteigt (KIRK und HONIG 1964, KREUZER und SCHOEPPE 1963).

In wieweit die aus diesem transmuralen Druckgradienten resultierende extravasale Widerstandskomponente die regionale Durchblutung des Myocards beeinflußt, ist nicht sicher geklärt. Die bisher vorliegenden Messungen haben widersprüchliche Ergebnisse geliefert; unbekannt ist bisher auch, in wieweit dieser Tatbestand sich auf die Durchblutung kollateral durchbluteter, von der normalen Versorgung ausgeschlossener Areale auswirkt. Diese beiden Fragen waren Gegenstand der folgenden Untersuchung.

Methodik. Die Untersuchungen wurden an 42 Hunden in Nembutalnarkose unter Kontrolle des arteriellen Blutdruckes, des HZV, der arteriellen O_2- und CO_2-Partialdrücke und des arteriellen pH durchgeführt. Zur Bestimmung der regionalen Myocarddurchblutung wurden zwei Verfahren alternierend eingesetzt:

Die Partikel-Verteilungsmethode (FLOHR 1970) sowie die Antipyrinmethode (REIVICH 1969). Mit beiden Verfahren wird die Durchblutung über die Aufnahme eines radioaktiven Indikators, dessen Verteilung autoradiographisch dargestellt wird, gemessen.

In einer ersten Serie (12 Versuchstiere) wurde die regionale Durchblutung des normal schlagenden Herzens untersucht. Blutdruck, Frequenz, HZV, PaO_2, $PaCO_2$ und arterieller pH dieser Gruppe lagen im Normbereich.

In einer zweiten Serie (30 Versuchstiere) wurde die regionale Myocarddurchblutung nach akutem Verschluß des R. descendens ant. über eine in einer Voroperation angelegte Silberdrahtschlinge untersucht. Die Versuchstiere wurden nach der Indikatorinjektion durch KCl getötet.

Resultate

1)Die Durchblutungsverteilung im normal schlagenden Myocard erscheint relativ homogen. Ein Durchblutungsgradient in der Wand des linken Ventrikels, der der inhomogenen Verteilung der extravasalen Widerstandskomponente entsprechen würde, wird nicht beobachtet. Daraus ist zu folgern, daß die in den Autoradiogrammen sichtbar werdende normale Durchblutungsverteilung offenbar das Resultat einer präzisen lokalen Durchblutungsregulation ist. Die extravasale Komponente des Widerstandes wird offenbar durch eine dem Betrag nach gleich große umgekehrt gerichtete vasale Komponente des Widerstandes neutralisiert.

2)Nach Occlusion des R. desc. ant. findet sich in den Autoradiogrammen eine gut abgrenzbare, scharf abgesetzte Zone mit reduzierter Indikatoraufnahme. Die sich aus der Indikatoraufnahme ergebende Durchblutungsgröße, d. h. die kollaterale Versorgung liegt zwischen 1o und 6o % der in den nicht betroffenen Partien gemessenen Durchblutung.

Innerhalb dieses Areals reduzierter Durchblutung ist eine charakteristische Inhomogenität der Indikatorverteilung zu beobachten. In den Innenschichten wird regelmäßig die stärkste Reduktion der Indikatoraufnahme gemessen, während im äußeren Drittel der Wand eine relativ hohe Indikatorkonzentration vorliegt. Das heißt, die Durchblutungsverteilung im kollateral versorgten Myocard ist inhomogen und entspricht in ihrem Muster dem Gradienten der extravasalen Widerstandskomponente(KREUZER und SCHOEPPE). Infarzierte Bezirke sind also durch zwei Merkmale gekennzeichnet: eine mehr oder weniger ausgeprägte Reduktion der Durchblutung und eine Störung der Durchblutungsverteilung.

Literatur

1)FLOHR, H. and A. HOPPE: Pflügers Arch. 31o, 16-21 (1969).
2)KIRK, E. S. and C. R. HONIG: Amer. J. Physiol. 2o7, 661-668 (1964).
3)KREUZER, H. and W. SCHOEPPE: Pflügers Arch. ges. Physiol. 278, 181-198 (196o).
4)REIVICH, M., J. JEHLE, L. SOKOLOFF and S. S. KETY: J. appl. Physiol. 27, 296-3oo (1969).

Responses of Coronary Vessels to Adrenergic Stimuli

By F. M. Abboud
Department of Internal Medicine, University of Iowa, USA

Experiments were done on anesthetized dogs (chloralose-urethane) to characterize the pattern of responsisveness to adrenergic stimuli. The left cir-
cumfles coronary artery was perfused with blood at constant rate. Changes
in perfusion pressure of the coronary vessels reflected changes in coronary
vascular resistance. Responses to direct electrical nerve stimulation of the
cardiac sympathetic nerves, to intra-coronary injections of norepinephrine,
isoproterenol, epinephrine; to electrical stimulation of carotid sinus nerves;
and to stimulation of carotid chemoreceptors with nicotine and cyanide were
tested. The intravenous administration of Practolol (1-2 mg/kg) eliminated
changes in contractility which could result from these interventions and thus
minimized or eliminated changes in coronary vascular resistance in response
to changes in myocardial metabolism. The results indicated that 1) there is
a paucity of alpha adrenergic vasoconstrictor receptors in the coronary vessels as compared to other vascular beds; 2) the coronary vascular beta receptors are not homologous to the cardiac beta receptors and are responsive
to isoproterenol and epinephrine but not to norepinephrine nor to mose sympathetic nerve stimulation; 3) most of the dilator effect of isoproterenol represents activation of coronary and not cardiac beta receptors; 4) stimulation of baroreceptor nerves causes withdrawal of the sympathetic vasoconstrictor influence and minimal activation of cholinergic vagal vasodilator
fibers; and 5) stimulation of chemoreceptors caused significant activation of
cholinergic vagal vasodilator fibers.

These observations and other studies in different vascular beds emphasize
the uniqueness of the responsiveness of the coronary circulation to adrenergic interventions.

Zusammenfassung und Diskussion

KNIERIEM: Der Prozeß der Sklerosierung geht von der Intima und nicht so sehr von der Media aus. Primär sind subendotheliale, glatte Muskelzellen und je nach Art der Schädigung auch glatte Muskelzellen der Media beteiligt.

FLECKENSTEIN betont die Unterschiede der Genese der durch AT 10 und Vitamin D induzierten Sklerose einerseits und der durch Cholesterinfütterung induzierten Sklerose andererseits. Immer aber sei die glatte Muskulatur der Media beteiligt. Nekroseherde wurden nur in der Media beobachtet.

KNIERIEM weist darauf hin, daß beim Rind Verkalkungen nicht in proliferier ten Muskelzellen, sondern in der Nähe von neugebildeten Kollagen und ela- stischem Material zu finden sind.

FLECKENSTEIN hebt hervor, daß bei den durch AT 10- und Vitamin D indu- zierten Sklerosen die Verkalkung die Initialzündung für den ganzen Prozeß darstellt, wobei die Hemmung des Calciumeinstroms (z. B. durch Magnesium Kalium, evtl. in Kombination mit organischen Calcium-Antagonisten) den Zelltod verhindern soll. Dabei muß unterschieden werden zwischen Calcium- Antagonisten, die vor allem am Herzen wirken und anderen, die besser am Gefäß wirksam sind. So entkoppelt z. B. Nitroglycerin an der Gefäßmuskula- tur, nicht jedoch an der Herzmuskelfaser. Magnesium beeinflußt weniger (die bei Vorbehandlung mit Natriumphosphat erhöhte) Plasmaphosphatkonzentra- tion in der Umgebung der Zelle, sondern blockiert vor allem die Anhäufung von Calcium im Zellinnern.

ZÀZVORKA weist darauf hin, daß bei den Aktivitätsbestimmungen im experi- mentell erzeugten Infarkt der Einfluß des Kollateralflusses nicht ausgeschal- tet werden konnte, aber durch Verwendung von Schweinen (10 % Kollateral- fluß gegenüber Hunden mit 20-25 % Kollateralfluß) so niedrig wie möglich ge- halten wurde.

Zur Frage von FLOHR nach der Einstichtiefe sagte BENZING: Die Ionenver- änderungen in minderversorgten Myokardbezirken wurden mit Elektroden be- stimmt, die etwa 3-4 mm tief in das Myokard eingestochen waren. Bei gele- gentlich sehr oberflächlicher Lage (1 mm unter der Oberfläche) wurden je- doch kaum Änderungen bei Gefäßdrosselung gefunden. Untersuchungen über unterschiedliche Reaktionen in verschiedenen Tiefen wurden dabei nicht durch geführt.

Auf die Frage von BERNE nach der Geschwindigkeit der Reaktionen antwortet BENZING: Die auftretenden Änderungen der Ionenaktivität bei Drosselung ei- nes Koronargefäßes waren durchschnittlich erst nach etwa 15 Sekunden deut- lich zu beobachten, wobei zu beachten ist, daß es sich um Messungen im In- terstitium des Myokards und nicht um intracelluläre Messungen handelte. Zum anderen dürfte die etwas verzögerte Ansprechzeit der Elektroden mit zu der erst nach einigen Sekunden auftretenden Reaktion beitragen.

Während die Anstiege der H^+-Aktivität im Interstitium bei Drosselung mit der Anhäufung saurer Stoffwechselprodukte (u. a. Milchsäure) gut zu erklären sind, dürfte die bei Sauerstoffmangelatmung (10 % O_2 in N_2) beobachteten geringfügigen Verminderungen der H^+-Aktivität (0.05-0.1 pH-Einheiten) einmal auf die pH-Verschiebung infolge Verminderung des Sauerstoffangebotes und möglicherweise die geringere CO_2-Produktion zurückzuführen sein. Inwieweit es sich tatsächlich um eine Gewebshypoxidose handelte in diesen Versuchen, wurde nicht geprüft. Immerhin kam es oftmals schon während der Sauerstoffmangelatmung bei erhöhter Myokarddurchblutung zu einem langsamen Wiederanstieg der H^+-Aktivität, was auf eine langsam beginnende und den gesamten Organismus betreffende metabolische Acidose hinweist. Dieses Verhalten zeigte sich noch deutlicher nach Umschalten auf Normalatmung (kurzfristiger H^+-Aktivitätsanstieg etwa 0.05 pH-Einheiten über den Ausgangswert).

Chairmen: A. Fleckenstein, H. Benzing

Zur klinischen Anwendung der Argonmethode

Von H. W. Heiss, M. Tauchert, H. Sonntag, K. Kochsiek, H. J. Bretschneider
und L. A. Cott
Physiologisches Institut, Lehrstuhl I und Medizinische Klinik, Kardiologische
Abteilung der Universität Göttingen, Germany

Zur klinischen Anwendung der Argonmethode (BRETSCHNEIDER et al., 1966)
für die Durchblutungsmessung von Herz und Niere mußten folgende Probleme
gelöst werden:

1) Exakt definierter Beginn der Argonaufsättigung. Vor der Messung atmet
der Patient zur Gewöhnung an die Untersuchungssituation durch ein Ruben-
Ventil Raumluft. Zum vorgegebenen Zeitpunkt "Null", dem Aufsättigungsbe-
ginn, wird dann momentan auf ein Gasgemisch aus 79 Vol % Argon und 21 Vol
% Sauerstoff umgeschaltet und so über einen Vorratsbeutel, ein Zwischenstück
und das Ruben-Ventil ein halboffenes System hergestellt. Der wache Patient
atmet spontan, der anästhesierte wird unter URAS-Kontrolle beatmet (endex-
spiratorischer CO_2-Gehalt: 4-4, 5 Vol %).

2) Simultane Gewinnung der Blutproben. Simultan mit der ersten Inspiration
des Patienten aus dem Vorratsbeutel wird die Entnahmepumpe gestartet, die
nach der kontinuierlichen Entnahmetechnik arterielle und organvenöse Blutproben in gasdichte, heparinisierte 5 ml Glasspritzen aufzieht. Am Ende der
fünfminütigen Aufsättigungsphase wird die Pumpe durch Auskuppeln der An-
triebswelle ohne Verzögerung angehalten. Anschließend erfolgt sofort die
Entnahme des arteriellen und organvenösen Endwertes manuell und simultan.

3) Vertretbare Belastung des Patienten. Mit Hilfe der Seldinger-Technik können arterieller und organvenöser Katheter (Goodale-Lubin 7F) unter Bild-
wandlerkontrolle gelegt und mehrere Vergleichsmessungen an einem Patienten durchgeführt werden. Das erhöhte Risiko bei der Sondierung einer Organ-
arterie entfällt, da es ausreicht, den arteriellen Katheter - er muß aus methodischen Gründen den gleichen Totraum wie der organvenöse Katheter haben - in die Arteria femoralis einzuführen. Er dient zusätzlich zur Registrierung der arteriellen Drucke. Das Herzzeitvolumen wird mittels der Thermo-
dilutionsmethode bestimmt. Die Sauerstottsättigungen messen wir mit einem
CO-Oximeter. Die Patienten tolerieren das Argon-Sauerstoffgemisch be-
schwerdefrei. Eine Prämedikation ist nicht erforderlich.

4) Einfache Handhabung des Meßsystems. Sofern ein Bildwandler zur Verfügung steht und Gelegenheit zur Druckregistrierung gegeben ist, bleibt die Anwendung der Argonmethode nicht auf das Herzkatheterlabor beschränkt, sondern ist auch im Operationssaal, auf der Intensivstation und in anderen ge-
eigneten Räumen möglich. Die Kosten für die Einrichtung und Unterhaltung
des gaschromatographischen Analysensystems sind im Vergleich zu anderen
Methoden gering. Die Analysen können von einer technischen Assistentin nach
Einarbeitung ausgeführt werden.

Mit Unterstützung der Deutschen Forschungsgemeinschaft im Rahmen des
SFB 89 Kardiologie Göttingen.

In den vergangenen Jahren haben wir mit der Argonmethode an etwa 150 Patienten Koronar- oder Nierendurchblutung gemessen und folgende Ergebnisse erhalten ($\bar{x}$ ± SEM):

a) <u>Koronardurchblutung</u> (RAU et al., 1968/69; KOCHSIEK et al., 1970/71; TAUCHERT et al., 1971; HEISS et al., 1971): 1. <u>Normalwert</u> (n = 10): 82 ± 3 ml/min · 100 g; Koronarreserve: + 400 % (Faktor 5). 2. <u>Aorteninsuffizienzen</u> (n = 11; unterschiedlicher Schweregrad): 119 ± 8 ml/min · 100 g; Koronarreserve: + 89 %. 3. <u>Aortenstenosen</u> (n = 8; unterschiedlicher Schweregrad): 136 ± 12 ml/min · 100 g; Koronarreserve: + 135 %. 4. <u>Hypertrophische obstruktive Kardiomyopathie</u> (n = 8): 89 ± 10 ml/min · 100 g; Koronarreserve: + 264 %. 5. <u>Mitralstenosen</u> (unterschiedlicher Schweregrad) Vorhofflimmern (n = 15): 76 ± 6 ml/min · 100 g; Koronarreserve: + 67 %. Sinusrhythmus (n = 14): 82 ± 5 ml/min · 100 g; Koronarreserve: + 204 % (Abb. 1). 6. <u>Kombinierte Aorten-Mitralvitien</u> (n = 12): 119 ± 12 ml/min · 100 g; Koronarreserve: + 51 %.

Die Angaben der Durchblutungsgrößen beziehen sich auf Ruhebedingungen. Die Koronarreserve, das Verhältnis des koronaren Gefäßwiderstandes unter Ausgangsbedingungen zu dem unter maximaler Koronardilatation, wurde durch fraktionierte Injektion von Dipyridamol bis zu einer Dosis von 0,5 mg/kg i.v. ermittelt.

b) <u>Nierendurchblutung</u> (HEISS et al., 1970): Werte von 5 nierengesunden Patienten: 352 ± 52 ml/min · 100 g. Durch Injektion von 0,24 g Aminophyllin i.v. konnte dieser Ruhewert bei unverändertem Herzzeitvolumen um 100 % gesteigert werden.

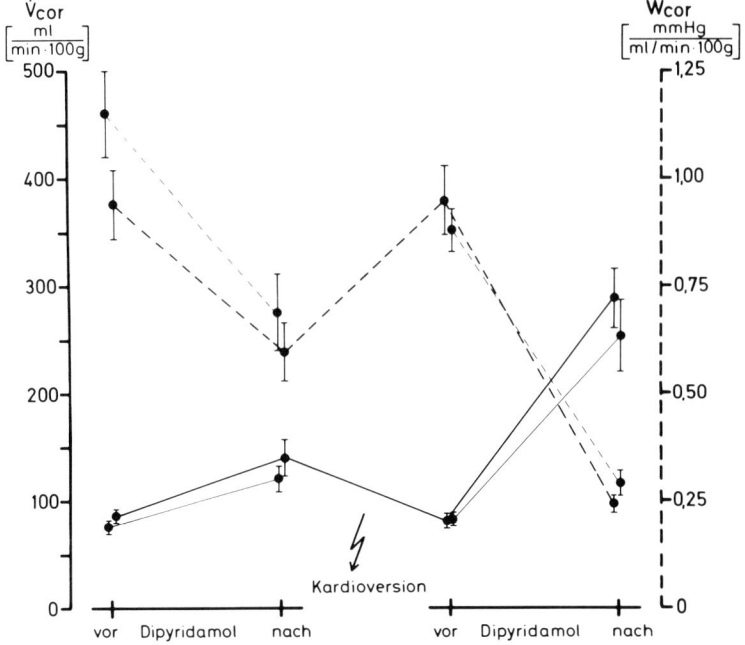

Koronardurchblutung (———) und Koronarwiderstand (– – – –) bei Patienten mit Mitralstenose. Links: unter Vorhofflimmern erheblich eingeschränkte Koronarreserve; rechts: Koronarreserve bei Sinusrhythmus deutlich verbessert. Dieser an zwei verschiedenen Kollektiven erhobene Befund wurde durch die Untersuchung einer dritten Gruppe (n = 10) bestätigt, bei der die Koronarreserve unter Vorhofflimmern und nach erfolgreicher Kardioversion gemessen wurde.

Theoretical VS Actual Indicator-Concentration-Time Curves in Parabolic Flow Model

By H. D. Green, C. E. Rapela, G. S. Malindzak, jr., and R. A. Gobbee
Bowman Gray School of Medicine, Winston-Salem, N. C. 27103, USA

Function of the capacitance vessels in a vascular bed (as contrasted to the resistance vessels) can be evaluated from the mean transit time ($\bar{T}$) and intravascular volume (Q), where $Q = \bar{T} \cdot \dot{Q}$ ($\dot{Q}$ = flow). $\bar{T}$ and $\dot{Q}$, theoretically, can be computed from indicator-concentration-time curves recorded at the outflow from a vascular bed following injection of an indicator into the inflowing stream. GONZALES-FERNANDEZ (GF) (Circulat. Res. 10, 409, 1962 - Eq. 4.14) derived an equation which describes the time course of indicator concentration for parabolic flow in a long straight tube. MALINDZAK et al. (BMES Absts., 1971) demonstrated that infinite time ($t = \infty$) is required to obtain correct computed values of $\bar{T}$ and $\dot{Q}$ from this relationship. At $t < \infty$ the computed $\bar{T}$ is < and $\dot{Q}$ > the correct values. In a physical model the observed appearance time (t_a) was correct but computed $\bar{T}$ was > and $\dot{Q}$ < the correct values, and the time to maximum concentration (t_{max}) was > that specified by the GF relationship. Correct values could be approximated only by limiting t to a value $<<\infty$. Preliminary studies suggest at least two causes for the discrepancy between theory and observation: a) "hang up" of the indicator at injection site due, in part, to the design of the injection site mixer (used to obtain flow tagging), and b) continuous washout of indicator from the injection port.

Aided by NIH grants HE-00487, -5392 and -00344 and North Carolina Heart Association.

Angiographische Darstellung von Koronararterienanomalien

Von J. Apitz

Abteilung für pädiatrische Kardiologie, Universitätskinderklinik, Tübingen, Germany

Die Koronararterienanomalien im weiteren Sinne sind in connatale und er-
worbene zu unterteilen.

Bei den connatalen Anomalien handelt es sich um Koronararterienfehlbildunge
Sie werden in Ursprungs-, Verzweigungs- und Verlaufsanomalien unterschieden. Isoliert vorkommend ist ein Teil dieser Fehlbildungen ohne größere
hämodynamische und klinische Bedeutung; als zusätzliche Anomalie bei einem angeborenen oder erworbenen Herzfehler können sie jedoch bei einem
operativen Eingriff zu ernsthaften Komplikationen führen. Eine präoperative
Kenntnis dieser begleitenden Koronararterienanomalien bei einem operablen
Herzfehler ist daher wünschenswert.

Ursprungsanomalien und anomale Verbindungen zwischen einer oder beiden
Koronararterien und einer Herzhöhle können bereits früh im Säuglingsalter
zu bedrohlichen Symptomen und zum Exitus letalis führen. Sie müssen daher
früh erfaßt, diagnostiziert und operiert werden.

Den erworbenen Koronararterienanomalien liegt eine Koronarsklerose zu-
grunde. Sie kann nodös oder diffus auftreten und zu einer Stenose am Abgang
oder im Verlauf einer oder beider Koronararterien bis zum völligen Verschluß führen. Die hierdurch bedingten Störungen der Myocarddurchblutung
führen zur Angina pectoris und u.U. zum Myocardinfarkt. Die Koronarsklerose ist die häufigste Erkrankung der Koronararterien. Sie hat in den letzten
Jahren sprunghaft zugenommen, ebenso wie die Zahl der Myocardinfarkte
und der koronar-bedingten Todesfälle. Seit einigen Jahren ist die Koronarsklerose eine der häufigsten Todesursachen im Erwachsenenalter. Dadurch
ist diese Erkrankung in den Mittelpunkt des Interesses der klinischen Cardio-
logie gerückt, zumal seit geraumer Zeit operative Maßnahmen zur Verfügung
stehen, mit denen diesen Patienten geholfen werden kann. Diese Operationen
erfordern eine genaue präoperative Diagnostik, bei der der Sitz, die Ausdehnung und der Schweregrad einer Koronarstenose bzw. eines völligen Ver-
schlusses dargestellt werden sollte.

Der sicherste Nachweis der einzelnen Koronararterienanomalien erfolgt
durch die Koronarographie. Die verschiedenen Methoden der Koronarographie werden besprochen und ihr Aussagewert für die einzelnen Anomalien
diskutiert.

Koronarstenosen und -verschluß. Angiographische tierexperimentelle Ergebnisse und Befunde beim Menschen

Von M. Thelen, R. Felix, N. Hahn, L. Beltz, D. W. Behrenbeck und H. Flohr
Radiologische Klinik, Chirurgische Klinik, Medizinische Klinik, Physiologisches Institut der Universität Bonn, Germany

Wenn auch die Koronardurchblutung im Koronarogramm nicht direkt meßbar ist, so kann man doch den Gefäßquerschnitt der extramuralen Koronararterien erfassen, der einen wichtigen regulierenden Faktor der Koronardurchblutung darstellt. Dabei ist von klinisch-radiologischer Seite zu unterstreichen, daß die zur koronaren Herzerkrankung führende Koronarsklerose eine Erkrankung der großen Hauptäste ist. Der Schwerpunkt der Ursachen, die zum Herzinfarkt führen, ist also in pathologischen Veränderungen der großen extramuralen Koronararterien zu suchen. Die tierexperimentellen angiographischen Untersuchungen, sowie die Bewertung menschlicher Koronarogramme gelten der Frage nach der Regulationsfähigkeit der extramuralen Koronararterien, dem Verhalten des Querschnitts hinter isolierten, koronarographisch dargestellten Stenosen und ob aus poststenotischen Querschnittsänderungen Rückschlüsse auf die Wirksamkeit einer Stenose möglich sind. Aus tierexperimentellen Untersuchungen ergibt sich dabei, daß nach plötzlichem Verschluß des Ramus interventricularis anterior die freien extramuralen Koronararterien keine spontane Querschnittsänderung, insbesondere keine kompensatorische Vasodilatation zeigt. Die Dilatationsfähigkeit kann dagegen im Hypoxieversuch und nach Gabe einer koronaraktiven Substanz nachgewiesen werden. Die Wirksamkeit des Verschlußes wird durch den gestörten Kontraktionsablauf des linken Ventrikels im Laevikardiogramm und durch den umschriebenen Durchblutungsausfall im Autoradiogramm belegt. Aufgrund dieser Untersuchungsergebnisse wurden menschliche Koronarogramme in einer etwa vergleichbaren Situation, unter besonderer Berücksichtigung der Koronarsklerose, auf das poststenotische Verhalten der extramuralen Koronararterien untersucht. Dabei ergibt sich, daß bei der haemodynamischen Bewertung einer extramuralen Stenose aus dem poststenotischen Gefäßverhalten unter Zugrundelegung einer lokalen hypoxischen metabolischen Regulation im Sinne einer Vasodilatation folgende Aussagen möglich sind.

1) Ein normaler poststenotischer Gefäßdurchmesser kann Zeichen einer haemodynamisch ineffektiven Stenose sein. Poststenotisch tritt keine Ischämie auf, sodaß ein Kompensationsmechanismus nicht beansprucht wird.
2) Ein normaler Gefäßverlauf kann aber auch bei haemodynamisch wirksamer Stenose vorliegen. Bei geänderten elastischen Wandeigenschaften des Gefäßes, als Antwort auf die poststenotische Hypoxie, ist der intravasale Druck bereits so stark erniedrigt, daß er den Gefäßquerschnitt über das ursprüngliche normale Maß hinaus nicht erweitern kann. Ohne metabolische Regulation wäre der Gefäßquerschnitt wahrscheinlich noch kleiner.
3) Eine poststenotische Vasodilatation, die ihre Ursache nicht in einer Koronarsklerose hat, ist Folge der Stenose, da die Querschnittszunahme als Regulation zur Kompensation der wirksamen Stenose im Sinne einer poststenotischen hypoxischen Vasodilatation aufzufassen ist.
4) Bei poststenotischer Vasokonstriktion ist die Wirksamkeit der Stenose si-

cher anzunehmen. Die Gefäßengstellung muß damit nicht unbedingt eine Koronarsklerose zur Ursache haben, sie kann auch funktionell erklärt werden.

Diskussion

Vortrag COTT

KALTENBACH:
Sie sprachen von der regionalen Myokarddurchblutung in dem Sinn, daß man die Durchblutung nach rechter und linker Coronararterie getrennt bestimmen kann. Die Xenonmethode hat eine weitere Anwendung, indem man in kleinen Bezirken messen kann und zwar dadurch, daß man das Xenon direkt in das Gefäß injiziert. Die Unterschiede zwischen rechter und linker Kranzarterien-durchblutung sind klinisch nicht sehr relevant, da es sich meist um viel kleinere Areale handelt, die von Störungen betroffen und unterperfundiert sind. Experimentell und intraoperativ kann man durch direktes Einbringen des Xenons in das Gefäß eben diese kleinen Bezirke gut erfassen. Damit wird die wirklich regionale Durchblutung, wie sie ja auch bei coronaren Durchblutungs-störungen des Menschen wichtig ist, erfaßt.

Gibt es beim Argon auch ein Isotop, das sich für die Anwendung beim Menschen eignet?

COTT:
Es gibt Isotopen, aber sie haben eine so kurze Halbwertszeit, daß sie in der Praxis nicht anwendbar sind.

HEISS:
Korrigieren Sie für das intravasale Volumen? Bei der Xenonmethode geht das intravasale Volumen mit ein.

KALTENBACH:
Wenn man in das Gewebe injiziert, dann entfällt das intravasale Volumen. Wenn man sowohl in das Gewebe als auch ins Gefäß injiziert, erhält man sehr gute Korrelationen. Schon deshalb scheint der Unterschied nicht sehr beträchtlich zu sein.

FLOHR:
Ist der Gewebeverteilungskoeffizient bestimmt?

COTT:
Für Argon ist er jetzt experimentell bestimmt und beträgt beim Myokard ohne Herzverfettung 1,1.

FLOHR:
Und für pathologisch verändertes Myokard?

COTT:
Da ist es nicht gemessen. Es kommt auf den Lipoidgehalt an, aber wegen der geringen Lipoidlöslichkeit von Argon im Gegensatz zu Xenon und Krypton sind die Schwankungen sehr viel geringer. Für Helium sind die Schwankungen noch geringer, weil hier noch eine geringere Lipoidlöslichkeit vorliegt. Der Verteilungskoeffizient des Heliums beträgt 1.

HEISS:
Der Verteilungskoeffizient von Argon zwischen einer Erythrozytenkonzentration und reinem Plasma beträgt 1 : 0,8, ist also sehr gering, sodaß die Methode relativ unabhängig vom Hämatokrit ist.

REIDEMEISTER:
Können Sie etwas über das Operationsrisiko sagen? Haben Sie aufgrund Ihrer Messungen Erfahrungen, wonach Sie eine Operation ablehnen?

Das Patientengut ist bisher noch zu gering, sodaß wir noch keine Korrelationen aufstellen konnten.

HEISS:
Bei 3 Patienten - es handelte sich um Aortenstenose IV. Schweregrades - war der Sauerstoffverbrauch 24 ml/100 g in Ruhe. Bei der deutlich eingeschränkten Coronarreserve (Faktor 1,2) kamen sie schon während der Operation ad exitum. Die Herzen waren nicht wieder belebbar.

Vortrag GREEN

Dr. DOW:
When you carried out your experiments did you use a commercially available densitometer since they have a mixer at the beginning of the curvette?

Dr. GREEN:
No, we constructed our own densitometer for the studies.

Dr. ABBOUD:
According to your presentation, there is a significant amount of error using the dye dilution technique in the traditional manner. Will you care to estimate what the actual error is?

Dr. GREEN:
If there is a mixer in the system the error will probably be about 150 %. Thi will increase to 300 % if no mixer is used.

Dr. DOW:
We have found, that when injections are made into the vena cava there is muc greater variability than when the injections are made directly into the right ventricle.

Dr. GREEN:
As you suggest, the ventricle must certainly act as a mixer for the marker.

Vortrag APITZ:

KNIERIEM:
Zunächst wollte ich kurz von einem Beispiel einer kombinierten, kongenitalen und erworbenen Coronaranomalie berichten bei einem 46-jährigen Mann mit Aortenklappenstenose bei verkalkter Endocarditis, wobei die Klappen re seziert wurden und durch ein Kugelventil ersetzt wurden. Bei dieser Operation wurde schon beobachtet, daß nur das rechte Coronarostium offen war.

Der Mann verstarb 5 Tage später und bei der Obduktion fanden wir tatsächlich nur die rechte Coronararterie und von der linken lediglich den ramus descendens und den ramus circumflexus. Bei der genauen Präparation fanden wir dann zunächst eine Verzweigungsanomalie, indem unmittelbar hinter dem Ostium der rechten ein Ast ungerhalb der Pulmonalklappe verlief und Anschluß fand an den ramus descendens der linken. Wir hatten aber das unbestimmte Gefühl, daß evtl. der Hauptstamm der linken Coronararterie vielleicht doch angelegt sein könnte und wir haben deshalb den ganzen Block in Serie geschnitten, d. h. im Aortenwurzelbereich und es fand sich dann tatsächlich eine sehr schmale, regelrecht angelegte linke Coronararterie, die aber im Ostiumbereich durch alte organisierte Thromben verschlossen war. Wahrscheinlich als Folge der durchgemachten Endocarditis, aber das war makroskopisch nicht erkennbar. Die Arterie war völlig glatt und endothelealisiert und erst eben durch die Serienschnittuntersuchung belegbar. Dies war eine Kombination einer erworbenen Anomalie mit einer Aortenstenose.

APITZ:
Sicher neigt ein großer Teil der Verlaufsanomalien später zu Coronarsklerose. Ist bei Ihrem Patienten, Herr KNIERIEM, der ramus descendens der linken Arteria coronaria aus der rechten Arteria coronaria entsprungen?

KNIERIEM:
Dieser Ast dieser Verzweigungsanomalie mündete etwa in Höhe auch des Abgangs zum ramus circumflexus der linken Kranzarterie.

KALTENBACH:
Sie haben wiederholt gesagt, daß kongenitale Anomalien sehr häufig zur Erwachsenencoronarsklerose führen. Von der Erwachsenenpathologie her kann man das eigentlich nicht bestätigen. Bei 300 Coronarangiographien sahen wir 2 Anomalien: Einen Patienten mit einer Fistel von der rechten Kranzarterie in den rechten Ventrikel und einen Patienten mit getrenntem Abgang von ramus interventricularis und ramus circumflexus, aber von einer Häufung der Anomalien kann eigentlich nicht die Rede sein.

Vortrag THELEN

KNIERIEM:
In wieviel Fällen haben Sie bei diesen Patienten mit schwerer Coronarsklerose eine Hypertonie beobachtet? Bei unseren postmortalen Coronarogrammen, die wir quantitativ ausgewertet haben, jedenfalls morphologisch quantitativ, ließ sich bei den Fällen mit Hypertonie eine deutliche Progredienz der Coronarsklerose bis in die Peripherie verfolgen.

THELEN:
Wir coronarographieren die Patienten nur bis zu einem systolischen Druck von 160 mmHg.

KNIERIEM:
Nun haben Sie doch vielleicht einige Fälle, die eine Linkshypertrophie hatten und dann vielleicht ein dekompensiertes Hochdruckherz, wo dann der Druck wieder auf 160 abgesunken ist. Haben Sie solche Fälle hier dabeigehabt?

THELEN:
Nein, hierbei nicht. Wir haben eine Auswahl getroffen, um die Streubreite der Fehler möglichst gering zu halten.

Chairmen: J. P. Gilmore, G. Kissling

The Effect of the Coronary Perfusion Pressure on the Performance of the Heart, its Dependency on Intravascular Blood Volume and Coronary Vascular Resistance

By G. Arnold, C. Morgenstern, and W. Lochner
Physiologisches Institut der Universität Düsseldorf, Germany

In studies on isolated guinea pig hearts and in open chest dog hearts in situ we could show that the perfusion pressure in the coronary arteries influences the performance of the heart. Furthermore we could demonstrate that it is the change of the perfusion pressure itself and not the related coronary blood flow that is responsible for the changes in the heart performance (1).

An augmentation of the coronary perfusion pressure results in an increase of the intracoronary blood volume, even if the coronary blood flow is kept constant. The same is true for the coronary blood flow, coronary perfusion pressure remaining constant. With the change of the intracoronary blood volume a change of the geometry of the heart can be observed: The myocardial wall thickness and the outer diameter of the heart increase with rising coronary perfusion pressure, while the inner diameter of the left ventricle is reduced (2).

Furthermore an enhancement of the coronary perfusion pressure does not change the force of contraction in case of an increased intracoronary blood volume due to a maximal coronary dilatation. Coronary vasoconstriction - induced by application of vasopressin and angiotensin - augments the effect of the coronary perfusion pressure.

The effect of the coronary perfusion pressure on the performance of the heart can be explained by an increased tension of the muscle fibres due to the augmentation of the intracoronary blood volume. This additional intracoronary blood volume caused by the increased pressure must be located in the prearteriolar part of the coronary vascular system.

ß-receptor-blocking agents do not effect the increase of the heart performance followed by an augmentation of the coronary perfusion pressure.

References

1) ARNOLD, G., MORGENSTERN, C., LOCHNER, W., OSWALD, S.: The autoregulation of the heart work by the coronary perfusion pressure. Pflügers Arch. $\underline{321}$, 34 (1970).
2) MORGENSTERN, C., ARNOLD, G., HÖLJES, U., WINDRATH, H.J., LOCHNER, W.: Der Einfluß des coronaren Perfusionsdruckes auf das intracoronare Blutvolumen und auf die Größe des linken Ventrikels. Pflügers Arch. $\underline{316}$, R16 (1970).

Myocard-Stoffwechsel und dessen Einfluß auf die Coronargefäßmuskulatur

Von R. Kadatz
Firma Dr. Karl Thomae GmbH. Biberach/Riß, Germany

Im Verhältnis zu seinem hohen O_2-Verbrauch ist die Durchblutung des Herzmuskels niedrig. Daraus resultieren eine hohe av-O_2-Differenz und ein niedriger venöser O_2-Gehalt. Änderungen des O_2-Verbrauches des Herzens bewirken sofortige Anpassung der Coronardurchblutung, wobei in weiten Grenzen von Ruhe und Belastung die O_2-Extraktionen und der venöse O_2-Gehalt von etwa 5 Vol % konstant bleiben. Der Stoffwechsel des Arbeitsmyokards beeinflußt offensichtlich die Weite der Coronargefäße, jedoch ist bis heute nicht sicher bekannt, wie diese Regulation erfolgt. Ein direkter Einfluß des myokardialen Sauerstoffdruckes auf die glatte Gefäßmuskulatur der Arteriolen ist unwahrscheinlich, auch der pCO_2, Kaliumfreisetzung sowie neurale oder neurohumorale Faktoren scheinen nicht entscheidend beteiligt zu sein. Nach der Hypothese von BERNE steuert das im Herzmuskelstoffwechsel gebildete Adenosin die Coronardurchblutung. Adenosin wirkt schon in geringer Konzentration coronarerweiternd und wurde zusammen mit seinen Abbauprodukten Hypoxanthin und Inosin während einer reaktiven Hyperämie im corona venösen Blut in Konzentration nachgewiesen, die bei intraarterieller Zufuhr für eine maximale Coronardilatation ausreichen. Beim normal mit O_2-versorgten Herzen wurde Adenosin auch im Herzbeutel gefunden. Es dürfte aus dem Extrazellulärraum des Myokards stammen, seine Konzentration ist höhe als im arteriellen Blut und steigt bei Hypoxie erheblich an. Adenosin entsteh im Myokard metabolisch durch Spaltung von ATP. Dieser Vorgang ist bei Zu nahme des O_2-Verbrauchs des Herzens gesteigert. Es permeiert frei durch Zellmembranen und gelangt aus dem Extrazellulärraum durch Diffusion zu den Gefäßen. Die resultierende Vasodilatation mit Durchblutungszunahme bewirkt einen beschleunigten Abtransport und enzymatischen Abbau des Adenosin. Diese Autoregulation läßt die Durchblutung wieder zu ihrem Ruhewert zurückkehren, sobald der durch erhöhten O_2-Verbrauch gesteigerte ATP-Zerfall sich wieder normalisiert. Als weitere metabolische Anpassung der Coronargefäße an einen erhöhten O_2-Bedarf wird von HONIG und Mitarb. die Neueröffnung von in Ruhe nicht durchbluteten Capillaren diskutiert. Die dadurch bewirkte Verkürzung der Diffusionsstrecke würde eine beträchtliche Verbesserung der O_2-Versorgung der Myokardfasern bringen.

Medikamentöse und chirurgische Therapie der koronaren Herzkrankheit im Zeitalter der selektiven Koronarangiographie

Von M. Kaltenbach, H.-J. Becker und G. Kober
Zentrum der Inneren Medizin, Sektion Kardiologie der Universität
Frankfurt a. M., Germany

Die Koronarangiographie hat die Voraussetzungen für eine intravitale morphologische Beurteilung der Herzkranzgefäße des Menschen geschaffen. Anhand der Befunde von über 250 Patienten mit selektiver Angiographie beider Koronararterien und Linksventrikulographie wird gezeigt, welche Fälle heute einer chirurgischen Therapie zugänglich sind. Die postoperativen Befunde und Verlaufsbeobachtungen der bisher operierten Patienten (VINEBERG-Operation, aortokoronarer Venen-Bypass und Aneurysmektomie) werden im Vergleich mit einem nichtoperierten Kollektiv dargelegt.

Die rationelle medikamentöse Therapie hat eine Prüfung der angewandten Substanzen bei der koronaren Herzkrankheit des Menschen zur Voraussetzung. Objektive Prüfungen koronarwirksamer Medikamente sind möglich, wenn die Koronarerkrankung angiographisch gesichert ist und wenn das EKG im Belastungsversuch eine reproduzierbare Ischämiereaktion erkennen läßt. Es wird über die Prüfung von Nitroglycerin, fünf verschiedenen langwirksamen Nitroderivaten, sechs Koronardilatatoren und fünf ß-Rezeptorenblockern berichtet; neben wirksamen fand sich auch eine relativ große Anzahl von unwirksamen Medikamenten, die keinerlei Effekt auf die Angina pectoris des Menschen objektivieren ließen.

Aufgrund eigener Erfahrungen und unter Berücksichtigung der Ergebnisse anderer Autoren wird versucht, den heutigen Standort einer sinnvollen medikamentösen und chirurgischen Therapie der koronaren Herzkrankheit des Menschen zu umreißen.

Koronarinsuffizienz und suspekter Diabetes

Von G. Kaiser
Kreiskrankenhaus Marktheidenfeld, Germany

Die Sklerose des Kranzgefäßsystems spielt die wichtigste Rolle in der Äthiopathogenese der Koronarinsuffizienz. Die vorzeitige Entwicklung der Koronarsklerose wird durch den Diabetes mellitus begünstigt. Typische diabetische Gefäßveränderungen schreiten zeitlich wesentlich dem manifesten Diabetes voran. Zur Früherkennung des Diabetes werden verschiedene Testmethoden empfohlen. Der Cortison-Glukose-Toleranztest nach Fajans und Cohn erwies sich als eine der empfindlichsten Provokationsmethoden.

In der vorliegenden Studie haben wir eine Gruppe von familiär nicht diabetesbelasteten Kranken mit Koronarinsuffizienz (77 Männer und 28 Frauen) auf einen suspekten Diabetes hin untersucht. Ausgeschloßen wurden Diabetiker, Übergewichtige und Patienten mit Erkrankungen, die einen Einfluß auf den Kohlenhydrat- bzw. Fettstoffwechsel haben.

Die Diagnose war elektrokardiographisch, biochemisch und durch die typischen klinischen Symptome gesichert. Als Diabetessuchverfahren wurde der Cortison-Glukose-Toleranztest nach Fajans angewandt.

Aufgrund dieses Verfahrens weisen in unserem Patientenkollektiv 68 % der Männer und 82 % der Frauen eine Störung der Glukose-Homöostase im Sinne eines suspekten Diabetes auf. Die Mehrheit der Kranken mit pathologischer Belastungskurve befindet sich in der Altersgruppe von 50-59 Jahren (42 % der Fälle). Der Nüchternblutzucker war in den einzelnen Fällen nicht wesentlich erhöht (n = 105; $\bar{x}$ = 109,6 mg %, s = $\pm$ 31,4), trotzdem lag der Mittelwert der Koronarpatienten höher als bei der entsprechenden Kontrollgruppe (n = 20; $\bar{x}$ = 88,2 mg %, s = $\pm$ 18,8). Der Unterschied ist statistisch signifikant (n = 105; t = 4,10; p < 0,001). Nach 12 Monaten haben wir 25 Patienten mit gestörtem Kohlenhydratstoffwechsel einer Kontrolluntersuchung unterzogen. Eine Gruppe wurde täglich mit Tolbutamid behandelt, eine andere nur mit einer entsprechenden Diät. Unsere Kontrollgruppe ist zu klein, um eindeutige Schlüsse zu ziehen. Viele Autoren vertreten jedoch die Meinung, daß die Glukoseintoleranz reversibel ist. Durch die optimale Stoffwechseleinstellung des Diabetes kann das Auftreten der Angiopathien verzögert und das Ausmaß abgeschwächt werden.

Definition des suspekten Diabetes (früher latenten):
1. Nüchtern - Blutzucker - Konzentration - normal
2. Orale oder intravenöse Glukose-Belastung - normal
3. Temporäre Kohlenhydratintoleranz (Schwangerschaft, Übergewichtigkeit, "Stress", Medikamentengebrauch, wie Corticosteroide, Thiazide usw.).

Anhand unserer Ergebnisse wurde erneut die hohe Koinzidenz von Koronarinsuffizienz und pathologischen Abweichungen des Kohlenhydratstoffwechsels
bestätigt. Deshalb muß bei den ersten klinischen Zeichen einer Koronarerkrankung nach dem Diabetes gefahndet werden und die Frühtestung in die Rou-
tinediagnose eingebaut werden.

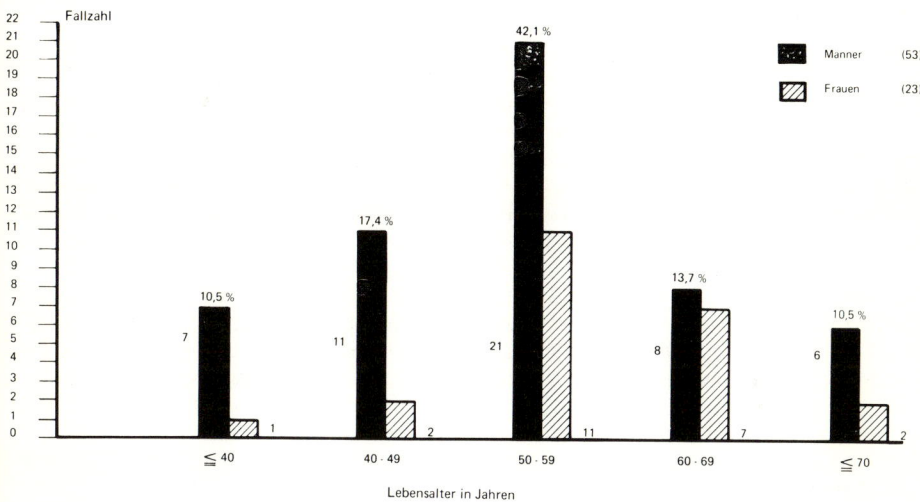

Altersverteilung der Patienten mit gestörtem Kohlenhydratstoffwechsel bei
Koronarinsuffizienz

Literatur

1) TEUSCHER A. u.a.: Neue Schweizerische Richtlinien zur Diagnose des
   Diabetes mellitus. Schweiz. med. Wschr. 101, 345 (1971).

Experimentelle Untersuchungen zu der Membran-Permeabilität von Cardiaca (g-Strophanthin, Digoxin und Ildamen) unter besonderer Berücksichtigung der Mikro-Autoradiographie

Von E. Löhr und H.-Br. Makoski
Röntgeninstitut II der Strahlenklinik des Klinikum Essen, Germany

Es wird über die Ergebnisse von Perfusionsversuchen an LANGENDORFF-Herzen berichtet, die mit den zwei Glykosiden g-Strophanthin und Digoxin sowie mit dem ß-Amino-Keton-Oxyfedrin (Ildamen) durchströmt wurden. Die Pharmaka waren mit C^{14} und H^3 markiert. Es wurde in vergleichenden Serien der zeitliche Ablauf hinsichtlich der Lokalisation dieser Cardiaca im extra- und intrazellulären Raum festgehalten.

Es zeigt sich, daß sowohl Digoxin und auch Ildamen vom Extrazellulärraum in breiter Front bereits 15 Minuten nach Beginn der Applikation nach intrazellulär vordringt. Besonders für Ildamen wird eine Affinität zu den Zellkernstrukturen beobachtet. g-Strophanthin ist vorwiegend extrazellulär gelagert. Es besteht demnach bei ähnlicher Glykosid-Wirkung bei g-Strophanthin und Digoxin ein differierender Befund hinsichtlich der Ablagerung dieser Pharmaka und deren Beziehung zu den Membranstrukturen, während für Digoxin und Ildamen ein ähnliches extra- und intrazelluläres Verteilungsmuster vorliegt.

Discussion

Paper Dr. ARNOLD

REIDEMEISTER:
Was there not a great deal of myocardial edema since the wall thickness increased about 15 %?

Answer: Yes, there was in the saline perfused hearts but not in those perfused with blood. We measured wall thickness by an ultrasonic method.

BALOURDAS:
Did you determine cardiac efficiency? What if any was the effect of decreased efficiency on coronary perfusion pressure?

Answer: The contractility was the parameter.

BALOURDAS:
Is there any relationship of coronary perfusion pressure to the normal or high systemic arterial pressure in your system and how does coronary perfusion pressure affect coronary sinus blood flow?

Answer: In the lower parts of coronary perfusion pressure the effects are more visible than in the higher parts.

Paper Dr. KADATZ

BERNE:
Dr. KADATZ has addressed some interesting questions to me at the end of his presentation. With respect to the coronary dilator effect of a rapid injection of 0.3 ml of venous blood into the coronary inflow tubing, I would not attribute this to myocardial ischemia and adenosine release but to ATP release from the red cells that were injected. I believe you will obtain the same dilation with the injection of arterial blood and it is probably due to disruption of the cells when they are rapidly injected through a hypodermic needle.

The potentiation of the coronary vasodilator action of adenosine by agents like dipyridamole is well accepted but the effects of these substances on reactive hyperemia in the heart is controversial; some investigators have reported potentiation whereas others have not. Actually, I would not expect to observe potentiation of reactive hyperemia with dipyridamole because it appears that the myocardial adenosine is chiefly in the interstitial fluid, where there are no degradative enzymes present to protect against. In the case of exogenous adenosine, the vasodilator effect is potentiated because the dipyridamole prevents the adenosine from entering the erythrocytes where it is deaminated to inosine.

Attenuation of the vasodilator effect of exogenous adenosine by theophylline (or aminophylline) attenuates reactive hyperemia is also controversial. In our experience aminophylline produced effects on reactive hyperemia that were quite similar to those observed with intracoronary adenosine administration, namely a decrease in the duration of the reactive hyperemia. Prof.

GERLACH has made similar observations in perfused hearts particularly when he increased the pH of the perfusion medium to compensate for the decrease in pH associated with the myocardial ischemia.

What are really needed in the studies with dipyridamole and aminophylline are measurements of tissue and venous effluent adenosine concentrations to provide a solid basis for interpretation of the effects of these drugs.

BETZ:
One has to discriminate between actions of intracoronary adenosine and the adenosine diffusing to intramyocardial vascular walls from hypoxic heart muscle cells.

In local tissue hypoxia of the myocardium the changes of potassium in the coronary sinus are very small compared with the extracellular potassium within the disturbed region. Therefore the coronary venous potassium concentration is only of minor significance for local myocardial hypoxia.

JACOB:
The ventricular myocardium is also supplied by cholinergic fibres. In these fibres Acetylcholine is released sufficient for causing a negative inotropic effect.

KAISER was asked: Do you have controls of the same type be answered that he did not have, but the tolerance for glucose decreases with age? This has to be taken in consideration during treatment.

LÖHR said that his perfusion pressure was 30 mmHg when asked by ARNOLD and that he did not see signs of underperfusion in the apical part of the heart.

Chairmen: R. M. Berne, G. Arnold

Zur Pathogenese der Coronarsklerose

Von G. Schmitt, U. St. Müller und W. H. Hauss
Medizinische Klinik und Poliklinik und Institut für Arterioskleroseforschung der
Universität Münster/Westf., Germany

Mittels Einbaukontrollen von ^{35}S-Sulfat in die Sulfomucopolysaccharide der
Grundsubstanz und ^{14}C-Hydroxyprolin in das Kollagen des Bindegewebes der
Gefäße haben wir in früheren Untersuchungen gezeigt, daß Mesenchymzellen
auf vielerlei Reize, auch und vor allem auf Risikofaktoren unmittelbar mit
einer Änderung ihres Stoffwechsels reagieren. Diese Erscheinung wurde als
"unspezifische Mesenchymreaktion" bezeichnet (zusammenfassende Darstellung: HAUSS, JUNGE-HÜLSING, GERLACH, 1968).

In tierexperimentellen Untersuchungen an Ratten wird nun gezeigt, daß nicht
nur der Leistungsstoffwechsel der Mesenchymzellen, sondern auch ihr Tei-
lungsstoffwechsel an dieser Reaktion beteiligt sind.

Wir haben im Tierexperiment an Ratten den Einfluß
1) einer arteriellen Blutdruckerhöhung, 2) einer zentralnervösen Reizung
durch Hypothalamus-Stimulation, 3) einer emotionalen Belastung und 4) toxischer Reize geprüft.

Bezüglich der Methoden zur Erzeugung eines akuten bzw. chronischen Hoch-
drucks, der Hypothalamus-Reizung, der emotionalen Belastung und der Toxin-
Injektion sei auf frühere Arbeiten verwiesen (5,4).

Bei allen Versuchstieren haben wir mittels ^{3}H-Thymidin und zwar als Vor-
und Nachmarkierung geprüft (1), ob die aufgezeigte Zellvermehrung infolge
mechanischer, nervöser und toxischer Reize in den Coronararterien durch
Reduplikation wandständiger Zellen oder durch eine Zelleinwanderung entsteht.

Befund: Vergleicht man die Coronararterie eines normotonen Tieres mit der
eines hypertonen (akute arterielle Hypertension durch Hypertensininfusion
bzw. chronischer experimenteller nephrogener Hochdruck) Tieres, so lassen
sich unter Hochdruckeinwirkung bei Nachmarkierung mit ^{3}H-Thymidin sowohl
im Bereich der Intima, der Media als auch der Adventitia erheblich mehr markierte Zellen nachweisen. Abb. 1 zeigt ein Beispiel: In der Coronararterien-
wand des normotonen Tieres (links) sind keine, in der Wand des hypertonen
Tieres dagegen viele Zellen markiert.

Die statistisch signifikante Erhöhung der Markierungsrate war nur bei Nachmarkierung nachweisbar, d.h. die arterielle Hypertension bewirkt eine Zell-
proliferation gefäßwandständiger Zellen. Bei Vormarkierung war dagegen keine über den Normbereich nachweisbare Steigerungsrate gegeben.

Hypothalamus-Stimulation, emotionale Belastung und Toxin-Injektionen bewirk-
ten eine erhöhte Markierungsrate nur bei Vormarkierung, nicht dagegen bei
Nachmarkierung. Diese Befunde weisen darauf hin, daß die nach Hypothala-

mus-Stimulation bzw. nach emotionaler Belastung und Toxin-Injektion nachweisbaren markierten Zellen dem hämopoetischen System entstammen (2).

Hierfür sprechen auch die Ergebnisse der Blutbildauszählung. Während man beim gesunden, nicht belasteten Tier bzw. beim belasteten Tier mit Nachmarkierung auf 100 mononukleäre Rundzellen nur 2-4 markierte Rundzellen im Blut antreffen kann, steigt die Markierungsrate nach Vormarkierung z.B. unter emotionaler Belastung bis auf 37 % an.

In früheren Untersuchungen (2,3) konnten wir bereits den Nachweis führen, daß die aus dem hämopoetischen System stammenden mononukleären Rundzellen in der Lage sind, sich zu transformieren und Bindegewebe zu bilden.

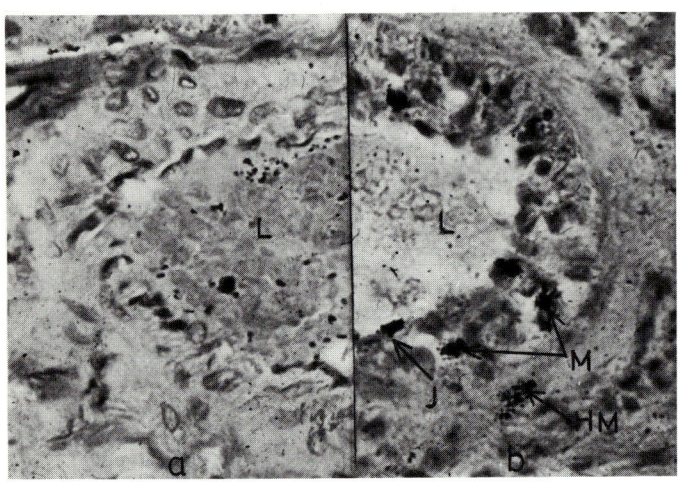

Abb. 1a. Ausschnitt einer intramuralen Coronararterie einer normotonen Ratte bei Nachmarkierung. Vergrößerung 400 : 1. - L = Lumen der Coronararterie. Keine markierten Zellen nachweisbar. Färbung H.E.
Abb. 1b. Ausschnitt einer intramuralen Coronararterie einer hypertonen Ratte bei Nachmarkierung. Vergrößerung 400 : 1. - L = Lumen der Coronararterie. I = markierte Intimazelle, M = markierte Mediazellen, H M = markierte Bindegewebszelle in der Herzmuskulatur. Färbung H.E.

Literatur

1) BÜCHNER, Th., JUNGE-HÜLSING, G., WAGNER, H., OBERWITTLER, W., HAUSS, W.H.: Klin. Wschr. 48, 467 (1970).
2) HAUSS, W.H., SCHMITT, G., MÜLLER, U.St., TILLMANN, P.: Med. Welt 22, 627-631 (1971).
3) HAUSS, W.H., SCHMITT, G., MÜLLER, U.St.: Verh. dtsch. Ges. inn. Med. (1971) (im Druck).

4) SCHMITT, G., KNOCHE, H., HAUSS, W.H.: Verh. dtsch. Ges. inn. Med. 336-341 (1970).
5) SCHMITT, G., HAUSS, W.H.: Angiologen-Kongress Bad-Nauheim 1970 (im Druck).
6) HAUSS, W.H., JUNGE-HÜLSING, G., GERLACH, U.: Die unspezifische Mesenchymreaktion. Stuttgart: Thieme 1968.

Intensivtherapie der coronaren Herzkrankheit

Von K. W. Schneider
Zentrum für Innere Medizin der Universität Würzburg, Germany

Die Letalität des Herzinfarktes hängt weitgehend von Zeitintervall zwischen Beginn der präcordialen Schmerzen und der Einleitung der Intensivtherapie ab. Die sogenannte Dunkelziffer der vor Behandlung verstorbenen Herzinfarkte beträgt nach neuesten Statistiken etwa 47 %. Hieraus ergibt sich die dringende Notwendigkeit der Klinikeinweisung bei länger als 15 Minuten anhaltenden infarktverdächtigen Symptomen. Die Hauptaufgabe einer coronaren Intensivpflegestation darf heute nicht mehr in der Behandlung des Herzstillstandes gesehen werden, sondern in der frühest möglichen Feststellung von prämonitorischen Symptomen und der Einleitung lebensentscheidender Präventivmaßnahmen der myocardialen elektrischen Instabilität. Das wichtigste therapeutische Rüstzeug der coronaren Intensivstation ist das Defibrillationsgerät. Die heute als überlegen zu betrachtenden Gleichstromapparate vermindern die Gefahr von Herzkammerflimmern. Im Gegensatz zu Wechselstromgeräten kann damit die gewünschte Dosis genau programmiert werden und Verbrennungen der Haut dadurch vermieden werden. Die beim Infarktkranken notwendige analgetische und sedative Therapie muß heute vor allem unter dem Gesichtspunkt unerwünschter Nebeneffekte auf Herzrhythmik und Ventilation erfolgen. Der Schwerpunkt der Intensivtherapie beruht auf der Prophylaxe und Therapie von Rhythmusstörungen. Hypoxämie und Alkalose spielen eine wesentliche Rolle bei der Entwicklung letaler Arrhythmien. Die Sauerstofftherapie beim Herzinfarkt wurde wegen der Abnahme des Herzminutenvolumens und Zunahme des peripheren Widerstandes kritisch beurteilt. Die Bedeutung der respiratorischen Alkalose und ihre mögliche Vermeidung durch rechtzeitigen Einsatz der Sauerstofftherapie wurde bisher unterschätzt. Beim akuten Myocardinfarkt ist bei folgenden Arrhythmien eine unmittelbare Cardioversion empfehlenswert: Vorhofflattern, Kammerflattern und Kammerflimmern, Kammertachycardie. Vorhoftachycardie und Vorhofflimmern stellen eine Indikation zur Cardioversion nur bei Herzfrequenzen über 120 oder

bei Hypotension dar. Die Behandlung der Sinustachycardie mit Digitalis ist
oft wenig sinnvoll. Die Erfolge mit gepaarter oder gekoppelter Stimulation
sind bis jetzt nur gering. Supraventrikuläre Extrasystolen sprechen meist
auf Chinidin an. Die Differenzierung zwischen digitalisinduzierter Vorhoftachycardie mit Block und Vorhofflattern ist beim Herzinfarkt von besonderer Bedeutung, weil die Cardioversion bei Vorhofflattern sicher wirksam ist,
während sie bei der Vorhoftachycardie gefährlich sein kann. Bei den ventri-
kulären Extrasystolen ist vor allem die sogenannte vulnerable Phase bzw.
das R- on T-Phänomen zu beachten. Bei Vorliegen einer Sinustachycardie
oder eines Herzblocks 2. bis 3. Grades sollte ein passagerer Schrittmacher
ohne Rücksicht auf die aktuelle Herzfrequenz gelegt werden. Seine Tätigkeit
tritt in Kraft, wenn die Frequenz unter 50/min sinkt. Neben den Rhythmusstörungen spielen nach wie vor die klassische Herzinsuffizienz und das Schock
syndrom bei der Intensivbehandlung der Coronarpatienten eine Rolle. Es ist
nicht nötig und unter Umständen sogar gefährlich, diskrete Symptome einer
Herzinsuffizienz beim Myocardinfarkt mit Digitalis zu behandeln, so z.B.
Galopprhythmus, Rasselgeräusche, Tachycardie. Trotz der besseren Über-
wachungsmöglichkeit sollte nicht vergessen werden, daß die Irritabilität des
Myocards gegenüber Digitalis beim Herzinfarkt sehr ausgeprägt ist und bereits kleine Dosen in den ersten Tagen Rhythmusstörungen auslösen können.
Bei der Behandlung des Schocksyndroms ist die Normalisierung des Blutvo-
lumens unter gleichzeitiger kontinuierlicher Kontrolle des Zentralvenendruck
nach wie vor der wichtigste Gesichtspunkt. Es konnte gezeigt werden, daß
Herzinfarktpatienten auch mit niedrigem Herzminutenvolumen nach Anwendun
von Volumenexpandern ihr Schlagvolumen vergrößern.

Frühmobilisation von Herzinfarktpatienten unter telemetrischer Kontrolle

Von D. Jeschke
Medizinische Klinik der Universität Tübingen, Germany

In der Rehabilitation von Herzinfarktkranken gehört die Übungsbehandlung zu
den wesentlichsten Maßnahmen. Uneinigkeit herrscht darüber, wann mit der
Mobilisation nach dem Infarktereignis begonnen werden soll. Der traditionel-
len Forderung, daß körperliche Belastungen einem Patienten erst nach einer
Phase der strengen Bettruhe von 4-6 Wochen zugemutet werden dürfen, steht
die Auffassung gegenüber, daß die Mobilisation sobald als möglich der cardialen Funktion angepaßt werden soll. Gestützt auf die günstigen Erfahrungen
vorwiegend des Auslandes (1, 2, 3, 4, 5, 6, 7, 8, 10, 11) hat 1968 die Weltgesund-
heitsorganisation (12) Richtlinien für die Frühmobilisation erarbeitet. Vor-

aussetzung dafür ist ein komplikationsloser Infarktverlauf. Symptome einer akuten cardialen Insuffizienz oder eines cardiogenen Schocks, maligne Herzrhythmusstörungen, schwere pektanginöse Schmerzzustände und Temperaturen über 39 $^{\circ}$C gelten als Kontraindikationen. Erst nach Besserung des klinischen Bildes ist die Mobilisation möglich. Die Übungsbehandlung ist der jeweiligen klinischen Situation und der individuellen Leistungsfähigkeit anzupassen und stufenweise zu steigern. Die körperlichen Belastungen dürfen keine Pulsfrequenzveränderungen von + 30/min / - 10/min gegenüber dem Ruhepuls, keine Herzrhythmus- oder Überleitungsstörungen, keine Atemnot, keine starke Ermüdung, keinerlei Zeichen einer cardialen Insuffizienz oder gar eines cardiogenen Schocks hervorrufen.

Die individuelle Anpassung der Übungsbelastung nach diesen Kriterien ist selbst für gut geschulte Krankengymnastinnen schwierig, wenn nicht gar unmöglich. Wir haben deshalb in unserer Klinik die Frühmobilisation, die am 3. Tag nach dem akuten Infarktereignis begann, unter telemetrischer Kontrolle des EKG durchgeführt. Die Patienten wurden zweimal täglich mit einem definierten, in der zeitlichen Dauer genau festgelegten, zwölfstufigen Übungsprogramm mit einer Gesamtdauer von 21 Tagen behandelt.

Ab Herbst 1970 wurde jeder Patient mit einem gesicherten Myocardinfarkt, wenn er das akute Infarktstadium überlebte, in dieses Programm aufgenommen. Es handelte sich um 44 Patienten mit einem Durchschnittsalter von 59,2 Jahren, unter denen 4 Frauen waren. Die Charakterisierung der Schwere des Myocardinfarktes durch den prognostischen Index von NORRIS und Mitarb. (9) ergab 4 Patienten mit einem prognostischen Index von< 4; 12 Patienten mit einem CPI von 4-5; 19 Patienten mit einem CPI von 6-7; 6 Patienten mit einem CPI von 8-9; 2 Patienten mit einem CPI von 10-11 und einen Patienten mit einem CPI von > 12. 3 Patienten mit einem CPI von 8-9 bzw. 10-11 verstarben während der Mobilisationsphase. Nur bei 18 der 44 Patienten bestanden keine Komplikationen während der akuten Infarktphase, sodaß sie am 3. bis 5. Tag nach dem Infarktereignis mobilisiert werden konnten. Bei den restlichen war die Übungsbehandlung innerhalb der ersten 14 Tage möglich.

Der Schweregrad eines Myocardinfarktes, charakterisiert durch den prognostischen Index, beeinflußte entscheidend die Mobilisation. Mit zunehmendem Schweregrad traten Komplikationen auf, die Verlängerungen auf einer Belastungsstufe, Rückstufungen oder beides erforderten. Dadurch ergab sich eine Verlängerung der durchschnittlichen stationären Behandlungsdauer, die bei den überlebenden 41 Patienten 38 Tage betrug.

Herzrhythmusstörungen wurden bei fast allen Patienten teils erst in höchsten Belastungsstufen beobachtet. Die Herzrhythmusstörungen wurden in leichte (vereinzelte supraventrikuläre oder ventrikuläre monotope Extrasystolen bis zu 5/min), mittelschwere (supraventrikuläre oder ventrikuläre Extrasystolen bis zu 10/min, als Bigeminus vorliegende Kupplungen mit einer Phasendauer bis zu 15 Sekunden, kurzzeitige supraventrikuläre Tachycardien und Überleitungsstörungen bis 2. Grades) und schwere (ventrikuläre Extrasystolen von mehr als 10/min, als Bigeminus, Trigeminus oder gar Quadrigeminus vorliegende gekoppelte Extrasystolen mit einer Phasendauer von über 15 Sekunden, polytope ventrikuläre Extrasystolen, ventrikuläre Tachycardien, "R auf T-Phänomen", Av-Überleitungsstörungen 3. Grades) unterteilt. Die Häufigkeit und der Schweregrad der Rhythmusstörungen nahm bei den prognostisch ernster zu bewertenden Patienten zu (Tab. 1). Gegenüber körperlichen Übungen im Liegen, Sitzen und Stehen führten orthostatische Belastungen nur in wenigen Fällen zu einer pathologischen Reizbildung.

Die telemetrische Kontrolle des EKG erwies sich als geeignete Methode, um bei Patienten mit frischem Myocardinfarkt eine frühzeitige, individuell angepaßte Übungsbehandlung gefahrlos durchzuführen und damit die Vorteile einer Frühmobilisation auf Physis und Psyche der Kranken auszunützen.

Rhythmusstörungen					
Schwer	1		3	4^{++}	2^{+}	1
Mittelschwer	1	4	8			
Leicht		6	6	2		
Keine	2	2	2			
	< 4	4-5	6-7	8-9	10-11	> 12

Prognostischer Index nach NORRIS et al.

Literatur

1) BRUMMER,P. et al.: Amer. Heart J. 52, 269 (1956).
2) BRUMMER, P. et al.: Amer. Heart J. 62, 478 (1961).
3) HAVIAR, V. et al.: Verh. dtsch. Ges. Kreisl.-Forsch. 34, 367 (1968).
4) HELLERSTEIN, H.K.: Circulation Suppl. 39, 40, IV-124 (1969).
5) KAINDL, F. et al.: Ann. Cardiol. Angéiol. 20, 147 (1971).
6) LEVINE, S.A. et al.: Lama 148, 1365 (1952).
7) MISSMAHL, H.P.: Fortschr. Med. 88, 401 (1970).
8) NÖCKER, J. et al.: Ärztl. Fortb. (1971).
9) NORRIS, R.M. et al.: Lancet 8, 274 (1969).
10) TAKKUNEN, J. et al.: Acta med. scand. 188, 103 (1970).
11) VARNAUSKAS, E. et al.: Läkartidningen 66, 2405 (1969).
12) World Health Organization, Regional Office Europe, Freiburg (1968).

Rehabilitation nach Herzmuskelinfarkt

Von U. Dembowski
Sanatorium Victoria der LVA Württemberg, Bad Nauheim, Germany

Die Rehabilitation wird bereits während des Krankenhausaufenthaltes eingeleitet, um dem Rekonvaleszenten die eigene Versorgung im häuslichen Bereich zu ermöglichen. Nach einer Konsolidierungsphase von 4-6 Monaten, die
der Patient zuhause verbringt, erfolgt die aktive Rehabilitation in einer Kurklinik. Seit einigen Jahren besteht die Tendenz, die Zwischenphase auf 1/4
Jahr oder weniger zu reduzieren, maßgebend sind auf der einen Seite psychologische Gründe, auf der anderen das Bestreben, frühzeitig das systemati-
sche Training aufzunehmen, um die Zeit der Arbeitsunfähigkeit abzukürzen.
Die bisherigen Ergebnisse, vorwiegend aus Schweden, USA und Israel sind
ermutigend.

Während der Rehabilitation in der Kurklinik stehen im diagnostischen Bereich
die Leistungsdiagnostik zur Dosierung des Trainings und die Kontrolle der
Risikofaktoren im Vordergrund. Die therapeutischen Maßnahmen konzentrie-
ren auf:
1) Die Fortsetzung der speziellen angiologisch-kardiologischen Behandlung
z.B. mit Antikoagulantien, Digitalisglykosiden und rhythmisierenden Medikamenten.
2) Das körperliche Training zur strukturellen Erweiterung der Kollateralen
im Myokard und zur Ökonomisierung der Herzarbeit.
3) Die Normalisierung der individuellen Risikofaktoren. Dabei kommt der
psychologischen Führung, z.B. in Fragen der Raucherentwöhnung, der Ge-
wichtsreduktion und der Diätbehandlung, besondere Bedeutung zu.
4) Einleitung spezieller Maßnahmen, z.B. einer operativen Therapie.

Körperliches Training und Gesundheitserziehung werden als Gruppentherapie
nach einem festen Programm durchgeführt. Z.Z. wird versucht, hierbei eine gewisse Standardisierung zu erreichen.- Die bisherige Form der Nachsor-
ge sollte durch Gruppenbehandlung am Heimatort - ähnlich den Anti-Coronary-
Clubs - mit gemeinsamem Training und psychologischer Führung ergänzt werden.

New Results Concerning the Mechanism of the Action of Nitroglycerol

By B. E. Strauer and H. Avenhaus
Department of Medicine, Division of Cardiology, University of Göttingen,
Germany

The action of nitroglycerol in angina pectoris is not yet clearly defined. Clas
sic theory attributes the action of nitroglycerol to an increased coronary
blood flow associated with an enhanced oxygen availability to the heart. On the
other hand, it has been suggested that nitroglycerol lowers myocardial oxyge:
requirements by reducing both cardiac work and myocardial contractility. Th
evaluation of the inotropic action, however, is complicated by the fact that
nitroglycerol produces a complex change and a rapidly fluctuating unsteady
state of hemodynamics (1). Therefore the direct inotropic action of nitroglycerol on human ventricular myocardium was studied in isolated human papillary muscles under controlled experimental conditions. The results demonstrate that nitroglycerol exerts positive inotropic effects on the isolated human ventricular myocardium.

Isotonic afterloaded contractions were employed in 7 isolated human left ventricular papillary muscles which have been surgically excised from patients
during mitral valve replacements. Signals of the extent of shortening, of tension development and their first time derivatives (velocity of shortening, velocity of relaxation, rate of tension development, rate of fall of tension) were
simultaneously recorded and displayed on an oscilloscope (2,3). Force-velocity relations were obtained by plotting the tension developed and the velocity
of shortening. Maximum velocity of shortening (V_{max}) has been calculated by
using HILL's equation. In each experiment nitroglycerol (glyceryl trinitrate)
was given to achieve final concentrations of $0,01$, $0,1$, $0,2$, $0,50$ and $1,0$ µg
ml. A five per cent alcoholic solution of nitroglycerol - not yet explosive -
has been used. At a constant preload, afterload and frequency of stimulation
nitroglycerol ($0,5$ µg/ml) increases the extent of shortening ($p < 0,001$), the
velocity of isotonic shortening ($p < 0,005$), the velocity of isotonic relaxation
($p < 0,005$), and the rate of tension development ($p < 0,01$). At a constant pre
load and frequency of stimulation force-velocity relations are shifted upward
with increases of both the preload velocity ($p < 0,005$) as well as V_{max} ($p <
0,005$) and the maximum isometric tension ($p < 0,005$). The increases of car
diac mechanics averaged about 22-28 per cent compared with control values.
The effects were dose-dependent in accordance with typical dose response
curves. The increases of both the velocity of shortening (preload velocity,
V_{max}) and the rate of tension development give evidence for a positive inotropic effect of nitroglycerol on the isolated human ventricular myocardium.

The question whether the inotropic effect can contribute to the beneficial action of nitroglycerol in angina pectoris must be closely related to the influence of the course of myocardial contraction and relaxation on coronary hem
dynamics. In angina pectoris the velocity of myocardial contraction and rela
xation is most commonly found to be reduced. Likewise, enddiastolic pressures and heart size are considerably elevated. Therefore, an increase of

the myocardial component of coronary vascular resistance is to be expected accompanied by a decrease of coronary blood flow and hence oxygen availability to the heart. In contrast, a positive inotropic intervention will lower the pathologically elevated myocardial component of coronary vascular resistance by improving the course of myocardial contraction and relaxation and by normalizing heart size and presumably wall tension. Consequently, coronary blood flow and oxygen availability to the heart will be normalized. Thus, it seems to be evident, that - with regard to the reductions of systolic and diastolic load factors of the heart - the effect of nitroglycerol on myocardial contractility can contribute to the relief in angina pectoris.

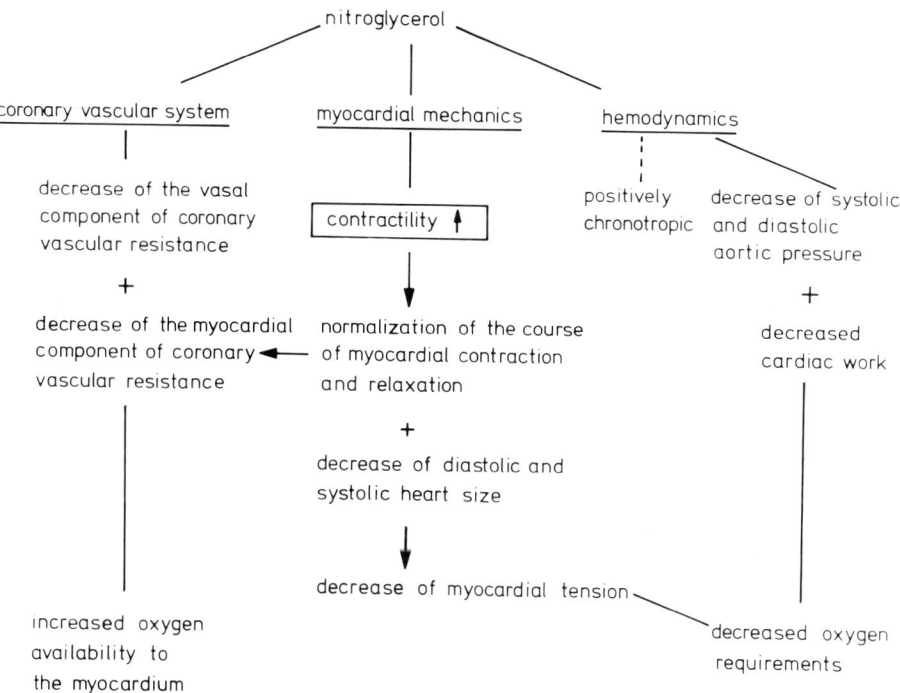

Schematic representation of the mode of action of nitroglycerol in angina pectoris, with special references to its effects on coronary circulation, myocardial contractility and hemodynamics. Nitroglycerol increases coronary blood flow by coronary vasodilatation (vasal component of coronary vascular resistance) as well as by a reduction of the myocardial component of coronary vascular resistance (improvement of the course of myocardial contraction and relaxation, decrease of myocardial tension). On the other hand cardiac oxygen requirements are reduced by decreases of myocardial tension, peripheral vascular resistance and cardiac work.

References

1) HONIG, C.R., TENNEY, S.M., GABEL, P.V.: Amer. J. Med. 29, 910 (1960).
2) STRAUER, B.E.: Klin. Wschr. 49, 468 (1971).
3) STRAUER, B.E., WESTBERG, C., TAUCHERT, M.: Pflügers Arch. 324, 124 (1971).

Der Einfluß der Acetylsalicylsäure auf die gesteigerte Thrombozytenaggregation bei dem Myokardinfarkt

Von I. Scharrer und K. Breddin
Abteilung für Angiologie des Zentrums der Inneren Medizin der Universität Frankfurt/Main, Germany

Schon 1965 berichtete OWREN über die vermehrte Plättchenadhäsivität bei Patienten mit Myokardinfarkten. Zahlreiche Untersucher fanden in den folgenden Jahren einen gesteigerten Plättchenumsatz, eine vermehrte Thrombozytenadhäsivität und eine erhöhte Verklumpungstendenz der Thrombozyten vor, während und nach dem Myokardinfarkt.

Wir konnten diese Befunde bestätigen. Bei folgenden Patientengruppen sahen wir eine signifikante Steigerung der Plättchenaggregation:
1) "Gefährdete" Patienten, die wir Wochen vor und kurz nach dem Herzinfark untersuchten
2) Patienten, bei denen der Herzinfarkt weniger als 6 Wochen zurücklag
3) Patienten, die vor mehr als 6 Wochen einen Herzinfarkt erlitten hatten.

Da in der Regel die Entstehung einer Thrombose mit der vermehrten Haftung und Aggregation von Thrombozyten an einer veränderten Gefäßwand beginnt, waren wir besonders interessiert, diesen Mechanismus zu hemmen, um damit möglicherweise die Infarkt- und Reinfarkt-Häufigkeit zu verringern.

Nach unseren und anderen experimentellen Untersuchungen ist die Acetylsalicylsäure der bisher wirksamste Aggregationshemmer in vivo. Wir beobachten zur Zeit etwa 50 Patienten mit Zustand nach Herzinfarkt, die als Reinfarktprophylaxe täglich 1,5 g Acetylsalicylsäure erhalten und weitere 50 gefährdete Patienten mit sogenannten Risikofaktoren, die unter einer Acetylsalicylsäureprophylaxe stehen. Die Ergebnisse dieser Studie werden diskutiert

Erlanger Erfahrungen mit der operativen Behandlung der koronaren Herzerkrankungen

Von J. von der Emde
Chirurgische Klinik der Universität Nürnberg-Erlangen, Germany

An der Chirurgischen Universitätsklinik Erlangen wurden seit 1967 bislang 170 Eingriffe zur Myokardrevaskularisation bei Patienten mit koronarer Herzerkrankung durchgeführt. Die Indikation zur Revaskularisation ist gegeben bei Patienten mit hochgradiger, für sogenannte Koronardilatatoren therapieresistenter Angina pectoris. Koronargefäßstenosen sind stets angiographisch lokalisiert. Vorausgesetzt wird die Einschränkung der Koronardurchblutung auf einer Seite von mindestens 80 %. Der aorto-koronare Bypaß wird bei segmentären Verschlüssen einzeln oder dreifach angelegt, sofern das Gefäß distal der Stenose ein Kaliber von über 2 mm hat. Liegen multiple Stenosierungen vor, sind insbesondere die kleineren Koronararterien befallen (Diabetes, Hypertonie) wird die indirekte Revaskularisation nach VINEBERG durch Implantation der A. mammaria in das hypoxische Myokard angewendet. Besteht eine latente Hypoxie des Myokards, d.h. ist das System nur durch verminderte Arbeitsleistung kompensiert, diffundiert Sauerstoff aus der implantierten A. mammaria dem Druckgefälle entsprechend in das Myokard. Nach Wochen bilden sich von der Adventitia ausgehend Kapillarsprosse, die sich weiter dehnen und ein ganz neues, jungfräuliches Gefäßsystem bilden. Es kommt zu einer retrograden Auffüllung des alten sklerotischen Koronargefäßsystems. Seit SONES durch die slektive Koronarangiographie nachweisen konnte, daß 80 % der Implantate bei adäquater Operationstechnik offen bleiben, war an dem Prinzip der VINEBERG'schen Operation nichts mehr zu bezweifeln. In der Literatur sind bisher 9 Fälle bekannt geworden, bei denen, autoptisch nachgewiesen, beide Koronararterien verschlossen waren und die Versorgung des Myokards nur noch über die implantierte A. mammaria lief. Unsere Operationsergebnisse werden statistisch dargestellt.

Zusammenfassung und Diskussion

Wenn nach der heutigen Ansicht die ischämische Herzerkrankung auf dem Boden einer Arteriosklerose entsteht und eine verschließende Koronarthrombo-
se auf dem Boden der Atherosklerose den hauptsächlichsten Faktor für einen
Herzinfarkt darstellt, so sind Untersuchungen von besonderem Interesse, die
a) die Pathogenese der Arteriosklerose und b) die Thromboseentstehung bzw.
-verhütung behandeln.

Aus dem HAUSS' schen Arbeitskreis (SCHMITT und Mitarb.), der sich intensiv mit der Pathogenese der Koronarsklerose beschäftigt, konnte neu in Tier-
versuchen gezeigt werden, daß zur initialen, obligaten Reaktion der Gefäßwand auf sklerogene Faktoren, der "unspezifischen Mesenchymreaktion",
auch Zellproliferationen kommen, die abhängig von sklerogenen Faktoren
sind. Hypertension führt zu einer Reduplikation gefäßwandständiger Zellen,
die anderen sklerogenen Faktoren führen zu Invasionen mononukleärer Zellen aus dem Blut. Grundsätzlich wird jedoch bei den Untersuchungen aus dem
HAUSS' schen Arbeitskreis die Frage zu erheben sein, inwieweit diese tierexperimentellen Ergebnisse auf die menschliche Pathogenese der Koronarsklerose übertragbar sind.

Es wird von K.W. SCHNEIDER die Frage aufgeworfen, ob die in den Tierex-
perimenten überprüften 4 Faktoren, insbesondere die arterielle Blutdruckerhöhung oder emotionelle Reize bei verschiedenen Rattenpopulationen in
gleicher Weise wirksam sind. Altersabhängigkeit wird diskutiert, ebenso wie
das Problem einer Potenzierung der Bindegewebseffekte bei gleichzeitiger
Einwirkung mehrerer Faktoren. Schließlich wird die Frage einer Beeinflussung der hypertensinogenen Reduplikationsrate durch Dauer der Hypertonie
bzw. antihypertensiven Maßnahmen gestreift.

Die als emotionelle Belastung geschilderten Reizmodelle bedürften einer Er-
gänzung, bis zu welchem Grad eine Trennung primär psychogener von primär hämodynamischen Auswirkungen möglich ist. Eine quantitative Beurteilung psychischer Noxen im Tierexperiment erscheint schwierig.

Die Bedeutung der vermehrten Thrombozytenadhäsivität und der erhöhten
Verklumpungstendenz der Thrombozyten vor, während und nach dem Myokardinfarkt unterstrich der Vortrag von I. SCHARRER und K. BREDDIN.
Nach den Untersuchungen erscheint ein gezielter Einsatz der Acetylsalicylsäure in der Prophylaxe und Therapie des Infarktes zweckmäßig und im Er-
folg der bisher geübten Antikoagulantientherapie vergleichbar, wenn nicht
gar überlegen zu sein.

Über die Behandlung des akuten Herzinfarktes (K.W. SCHNEIDER) und die
bereits in der Klinik einzuleitende Rehabilitation nach Herzinfarkt (D. JESCH
KE, U. DEMBOWSKI) bestehen keine wesentlich divergierenden Meinungen.
Es wird unterstrichen, daß ein Patient mit akutem Herzinfarkt ohne Ausnahme
baldmöglichst der klinischen Behandlung mit der Möglichkeit einer modernen
apparativen und laborchemischen Überwachung bedarf. Allein damit ist eine
differenzierte Therapie der Komplikationen durchführbar. Eine körperliche
Übungsbehandlung, die der klinischen Situation und der individuellen Leistungsfähigkeit Rechnung trägt, ist als wesentlichste Maßnahme in der Reha-

bilitation von Herzinfarktkranken anzusehen. Zu diskutieren ist, ob sich der Trainingseffekt bei diesen Patienten z. B. durch Hypoxiebedingungen steigern läßt. In der Frühphase des Infarktes bedarf aber die KG-Behandlung einer sorgfältigen Kontrolle. Die Frühmobilisation in der Klinik ist mit ein entscheidender Schritt, um in einem anschließenden Heilverfahren die Früh- und Vollrehabilitation anzustreben.

Zum Thema der Frühmobilisation von Herzinfarktpatienten wird die Frage aufgeworfen, wie der Begriff "unkomplizierter Infarkt" definiert werden kann und wann hierfür der richtige Zeitpunkt gegeben ist. Das Fehlen von Rhythmusstörungen bis zum dritten Tag scheint keine ausreichende Legitimation zum Beginn des Übungsprogrammes. Unerwartetes Kammerflimmern kann bis drei Wochen nach Infarktbeginn ohne Prodromalerscheinungen auftreten. Eine Verbesserung der Kollateralenbildung, wie sie durch körperliches Training erreicht wird, dürfte, so meint K. W. SCHNEIDER, in den ersten vier Wochen nach Infarkt, welche Maßnahmen auch immer ergriffen werden, kaum so weit realisierbar sein, daß sie einen echten Nutzen verspricht. Dem Prinzip der Frühmobilisation wird das Programm des sogenannten "prolonged bed rest" gegenübergestellt. Neben körperlichem Training und Fortführung spezieller medikamentöser Therapieformen ist die psychologische Führung und Schulung das wesentliche Anliegen im Kurkrankenhaus.

In der Diskussion zum Vortrag der Intensivtherapie der koronaren Herzkrankheiten wird von K. W. SCHNEIDER pointiert die Forderung vertreten, daß bei einem jährlichen Aufnahmegut von 25 Herzinfarkten die sogenannte "coronary care unit" von der allgemein Inneren Medizin getrennt werden muß. Die Nichtverselbständigung der kardiologischen Intensivstationen ist ein schwerwiegender Fehler, der vor allem bei der Organisation des Pflegedienstes und der Koordinierung pflegerischer Maßnahmen ins Gewicht fällt. Kardiale Intensivmedizin darf keine absolute Originalität beanspruchen. Zum Teil ist es sicherlich eine Folge des Mangels an Pflegekräften, die zu einer Zentralisierung von Patienten zwingt. Kardiologische Intensivmedizin wäre aber nicht in der heutigen Form ohne die technische Ausrüstung denkbar, die wiederum spezielle, personelle Voraussetzungen erforderlich macht. Die Einrichtung einer kardiologischen Intensivstation sollte nicht vor Sicherstellung der personellen Voraussetzungen gestattet sein. Die Intensivpflege beinhaltet Lagerung des Patienten, Sonderernährung, Messung der Ein- und Ausfuhr, aber auch die Überwachung bestimmter, elektronisch gesteuerter Geräte, die nicht nur vitale Funktionen des Patienten übernehmen, sondern auch kontrollieren. Das Pflegepersonal der Intensivstation muß also bei kardiologischen Fällen in der Lage sein, Tätigkeitsmerkmale festzustellen, die bisher ausschließlich von medizinisch-technischen Assistentinnen wahrgenommen wurden. Beobachtung der Herzstromkurve auf dem Bildschirm ersetzt nicht die Registrierung. Zahlreiche Verlaufskontrollen sind nur dann sinnvoll, wenn sie optimal protokolliert sind und jederzeit zur Verfügung stehen. Solange kein ausreichendes paramedizinisches Personal im Tag- und Nachtdienst zur Verfügung steht, müssen die skizzierten Aufgaben von Pflegepersonen übernommen werden. Eine Stationsoberschwester im alten Sinn, die allein ärztliche Weisungen entgegennimmt und die direkten Anordnungen des Arztes an die für bestimmte Patienten zuständigen Schwestern unterbindet, hindert nicht nur die optimale Versorgung und Überwachung der Patienten, sondern auch die berufliche Fortbildung.

Einen Beitrag zur medikamentösen Behandlung von ischämischen Herzer-
krankungen lieferte B. E. STRAUER, der an isolierten Papillarmuskeln des
Menschen die positiv inotrope Wirkung des Nitroglyzerins nachwies. Der
Beitrag ergänzt das bisherige empirische und theoretische Wissen über den
Wirkungsmechanismus von Nitritkörpern.

Chairmen: K. W. Schneider, D. Jeschke

Rundtischgespräch

Die koronare Herzkrankheit

Moderator: U. DEMBOWSKI
Teilnehmer: J. von der EMDE, G. JUNGE-HÜLSING, H.-J. KNIERIEM,
W. OBERWITTLER, K.W. SCHNEIDER

Das Gespräch wurde mit Fragen zur Pathogenese des Herzmuskelinfarktes
eingeleitet.

Der typische große Herzmuskelinfarkt ist die Folge einer vollständigen Ischämie des entsprechenden Gewebsbezirkes durch Verschluß einer Koronararterie im Bereich des Hauptstammes oder eines großen extramuralen Astes.
Als Ursache des Verschlußes findet man meistens eine Thrombose auf einem
arteriosklerotischen Beet. Der Verschluß kann aber auch durch ein akutes
Endothelödem, eine frische Blutung in die Gefäßwand oder eine Embolie entstehen. Bei sorgfältiger Untersuchung der Koronararterien mit Querschnit-
ten und nachfolgender histologischer Kontrolle wurden am Düsseldorfer Pathologischen Institut in 85 % der Infarkte thrombotische Verschlüsse gefunden.

Verschiedene Autoren wiesen in letzter Zeit darauf hin, daß in tabula bei einer Gruppe, die das Infarktereignis nur wenige Stunden überlebte, seltener
Thrombosen bestanden als bei einer anderen Gruppe, die den Infarkt mehrere Tage überlebt hatte. Diese Beobachtung könnte dahin gedeutet werden, daß
die Thrombose nicht Ursache, sondern Folge des Infarktes ist. Eine abschlies-
sende Stellungnahme zu dieser Frage ist noch nicht möglich.

Beim Schwielenherzen oder bei exzentrischer Linkshypertrophie kann eine
große Herzmuskelnekrose oder ein plötzliches Herzversagen auch durch eine
akute Koronarinsuffizienz ohne frische Thrombosierung eintreten.

Kleine Herzmuskelinfarkte entstehen gelegentlich durch Verschluß kleiner
intramural gelegener Koronararterienäste. Diese periphere Koronarsklerose
sieht man besonders bei Hypertonikern. - Disseminierte subendothelial ge-
legene Herzmuskelnekrosen sind auf eine Koronarinsuffizienz ohne vollständigen Verschluß zurückzuführen, d.h. sie sind Folge eines Mißverhältnisses
zwischen dem Blutbedarf des Myokards und dem durch die Koronarsklerose
eingeschränkten Blutangebot. Der Blutbedarf wird vor allem durch Erhöhung
der Herzschlagfrequenz und des Myokardstoffwechsels (Kathecholaminaus-
schüttung) gesteigert, das effektive Blutangebot kann durch einen kritischen
Blutdruckabfall, durch Erhöhung der Blutviskosität oder durch Erhöhung der
Herzschlagfrequenz vermindert werden. Zu diskutieren sind außerdem Störungen der Mikrozirkulation infolge eines Ödems der Kapillarendothelien.

Die zweite Themengruppe betraf die epidemiologischen Studien über den Herzinfarkt, die vor allem in den USA und in den skandinavischen Ländern durchgeführt wurden. Dabei ist zu betonen, daß die epidemiologische Arbeitsrichtung geeignet ist, Korrelationen aufzuzeigen, aber nur bedingt einen Hinweis
auf kausale Zusammenhänge geben kann. So ist z.B. die kausale Beziehung
der sogenannten Risikofaktoren zum Herzinfarkt völlig offen. Eher sind die

Risikofaktoren als Indikatoren für eine fortschreitende Arteriosklerose und für eine Gefährdung durch arteriosklerotische Komplikationen anzusehen. Am Beispiel der Framingham-Studie läßt sich zeigen, daß bei 30 % der Infarktpatienten kein einziger Risikofaktor, bei weiteren 41 % lediglich ein Risikofaktor vorgelegen hat.

Zunehmende praktische Bedeutung wird die epidemiologische Arbeitsrichtung zweifellos für die weitere Entwicklung der Präventiv-Medizin gewinnen. Voraussetzung für eine rationelle Arbeit auf diesem Gebiet ist der Einsatz maschineller Diagnosehilfen und spezieller statistischer Methoden.

Das weitere Gespräch wandte sich dann Fragen der Praxis und der Klinik zu, wobei zunächst die Aufgaben des erstbehandelnden Arztes, also des Hausarztes, diskutiert wurden. Die modernen Methoden der apparativen Überwachung und der Intensivpflege, die nur im Krankenhaus eingesetzt werden können, haben mit Recht zu der Forderung geführt, jeden Patienten mit dem Verdacht auf einen frischen Myokardinfarkt umgehend in ein entsprechend ausgerüstetes Krankenhaus einzuweisen. Gerade in den ersten Stunden ist nämlich die Gefährdung am größten. Leider vergeht oft wertvolle Zeit, bevor der Patient seinen bedrohlichen Zustand erkennt und den Hausarzt benachrichtigt. Es ist deshalb eine wichtige Aufgabe, gefährdete Personen in psychologisch geeigneter Weise auf die Hauptsymptome eines Infarktes und auf die Dringlichkeit der klinischen Behandlung hinzuweisen.

Trifft der Hausarzt den Patienten noch zu Hause an, sollte er bis zum Eintreffen eines Krankenwagens nach folgendem Schema verfahren:
1) Schmerzstillung z. B. mit 1 Amp. Fortral i. v. oder 1 Amp. Dolantin, bei leichten Schmerzen Valium i. v.
2) Bei latenter Herzinsuffizienz - diese ist in etwa 70 % der Fälle vorhanden - Digitalisierung mit 1/8 mg Digoxin i. v. Die Indikation kann hierbei großzügig gestellt werden.
3) Bei stärkerem Blutdruckabfall sind Kreislaufmittel angezeigt, z. B. Akrinor, das in diesen Fällen den Adrenalinabkömmlingen vorzuziehen ist. Außerdem gibt man 50 mg eines Cortisonpräparates i. v. Bei längeren Transportwegen kommt auch die Infusion eines Plasmaexpanders, z. B. Rheomacrodex, infrage.
4) Über den Einsatz von Antiarrhythmika sind die Auffassungen geteilt. Bei Extrasystolie kann man Xylocain in einer Dosis von 50-100 mg langsam intravenös injizieren.
5) Während des Transportes sind, wenn der Patient hierdurch nicht zu stark irritiert wird, Sauerstoffgaben zweckmäßig.
6) Die Heparinisierung erscheint nicht so dringend, der Hausarzt kann auf sie verzichten.

Da die Intensivpflege und die konservative klinische Behandlung bereits in den vorangegangenen Vorträgen dargestellt wurde, blieb der letzte Teil des Podiumsgespräches den operativen Behandlungsmethoden der therapieresistenten Angina pectoris vorbehalten. Die Revaskularisierung, die zunehmendes Interesse findet, setzt voraus, daß die Herzmuskelfasern erhalten und nicht bereits bindegewebig ersetzt sind. Dem Eingriff geht prinzipiell eine Koronarangiographie voraus. Bei isolierten Stenosen oder Verschlüssen des Hauptstammes der Koronararterie wird ein aortokoronarer Bypaß angelegt, bei

multiplen Stenosierungen, insbesonderer kleinerer Gefäßäste, empfiehlt sich
die indirekte Revaskularisation nach VINEBERG durch Implantation der A.
mammaria in den hypoxischen Myokardbezirk. Es ist zu erwarten, daß die
Zahl der gefäßplastischen Operationen am Herzen in nächster Zeit rasch an-
steigen wird. Bildet sich nach einem Myokardinfarkt ein Herzwandaneurysma
aus, so ist die Resektion des Aneurysmas angezeigt, dabei spielt seine Größe
eine untergeordnete Rolle. Es wurde auch schon versucht, bei frischem Infarkt die Resektion des infarzierten Gebietes vorzunehmen. Diese Methode
ist aber für eine breite Anwendung bisher noch nicht geeignet.

Subject Index